James Napier

CCEA GCSE
BIOLOGY QUESTIONS

COLOURPOINT EDUCATIONAL

© James Napier and Colourpoint Creative Ltd 2023

Print ISBN: 978 1 78073 188 9

First Edition
First Impression 2023

Layout and design: April Sky Design
Printed by: GPS Colour Graphics, Belfast

All rights reserved. No part of this publication may be reproduced, stored in a retrieval system or transmitted in any form or by any means, electronic, mechanical, photocopying, scanning, recording or otherwise, without the prior written permission of the copyright owners and publisher of this book.

Copyright has been acknowledged to the best of our ability. If there are any inadvertent errors or omissions, we shall be happy to correct them in any future editions.

The Author

James Napier is a former Vice-Principal in a large Northern Ireland grammar school. Dr Napier has written and co-written a number of GCSE Biology and Science textbooks supporting the work of teachers and students in Northern Ireland. He has also published a range of popular science books throughout the areas of genetics and evolution.

Colourpoint Educational
An imprint of Colourpoint Creative Ltd
Colourpoint House
Jubilee Business Park
21 Jubilee Road
Newtownards
County Down
Northern Ireland
BT23 4YH

Tel: 028 9182 0505
E-mail: sales@colourpoint.co.uk
Web site: www.colourpoint.co.uk

Publisher's Note: This book has been written to help students preparing for the GCSE Biology specification from CCEA, and the Biology units of the GCSE Double Award Science specification. While Colourpoint Educational and the author have taken every care in its production, we are not able to guarantee that the book is completely error-free. Additionally, while the book has been written to closely match the CCEA specification, it is the responsibility of each candidate to satisfy themselves that they have fully met the requirements of the CCEA specification prior to sitting an exam set by that body. For this reason, and because specifications change with time, we strongly advise every candidate to avail of a qualified teacher and to check the contents of the most recent specification for themselves prior to the exam. Colourpoint Creative Ltd therefore cannot be held responsible for any errors or omissions in this book or any consequences thereof.

Health and Safety: This book describes practical tasks or experiments that are either useful or required for the course. These must only be carried out in a school setting under the supervision of a qualified teacher. It is the responsibility of the school to ensure that students are provided with a safe environment in which to carry out the work. Where it is appropriate, they should consider reference to CLEAPPS.

CONTENTS

Unit 1
Cells, Living Processes and Biodiversity

1.1 Cells 5
1.2 Photosynthesis and Plants..7
1.3 Nutrition and Food Tests 11
1.4 Enzymes and Digestion..14
1.5 The Respiratory System, Breathing and Respiration..........................17
1.6 Nervous System and Hormones...20
1.7 Ecological Relationships and Energy Flow 27

Unit 2
Body Systems, Genetics, Microorganisms and Health

2.1 Osmosis and Plant Transport 37
2.2 The Circulatory System 40
2.3 Reproduction, Fertility and Contraception..43
2.4 Genome, Chromosomes, DNA and Genetics....................................47
2.5 Variation and Natural Selection 54
2.6 Health, Disease, Defence Mechanisms and Treatments 60

Note: This book is designed to be used by both Double Award Biology candidates and GCSE Biology candidates. Questions that should ONLY be attempted by GCSE Biology candidates are indicated with grey shading, as shown here. These questions should NOT be attempted by Double Award Biology candidates.

Note: Candidates will be in one of two tiers – Foundation Tier or Higher Tier. Questions that should ONLY be attempted by Higher Tier candidates are indicated with the words "HT ONLY" in the margin, as shown here. Foundation Tier candidates should NOT attempt these questions.

HT ONLY
HT ONLY
HT ONLY

Answers: The answers for this book are available online. Visit www.colourpointeducational.com and search for *Biology Questions for CCEA GCSE*. The page for this book will contain instructions for downloading the mark scheme. If you have any difficulties please contact Colourpoint – details on the previous page.

Unit 1
Cells, Living Processes and Biodiversity

1.1 Cells

1. Animals, plants and bacteria are formed of microscopic units called cells.
   (a) Name **two** cell components present in the cells of all three types of organisms. [2]
   (b) Name **two** parts of a cell present in plant cells only. [2]

2. The diagram below represents a cell from the leaf of a plant.

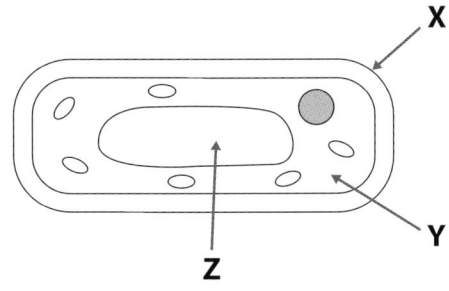

   (a) (i) Name the parts of the cell labelled **X, Y** and **Z**. [3]
       (ii) Which **one** of these parts is also present in animal cells? [1]
   (b) (i) In which part of the cell are mitochondria present? [1]
       (ii) Give the function of mitochondria. [1]
   (c) Name **two** structures which are present in bacterial cells but **not** in plant or animal cells. [2]

   HT ONLY

3. Describe how to prepare a slide of onion epidermal tissue for viewing under a light microscope. [4]

4. Draw and complete the table below about SI units.

Unit	Number in one metre	Standard form	Symbol
	1000	10^{-3} m	
micrometre			µm

   [4]

5. (a) A plant cell in a photograph is 4 cm long. The magnification of this cell is × 400. Calculate the actual size of the cell. [2]
   (b) The cell in another photograph is 70 mm. This cell has an actual size of 140 µm. Calculate the magnification of the cell. [2]

UNIT 1: CELLS, LIVING PROCESSES AND BIODIVERSITY

6. (a) Using the scale bar below, calculate the magnification of the cell shown. [2]

(b) Calculate the length of the cell. [2]

7. Electron microscopes have a much higher resolution than light microscopes. State what is meant by the term resolution. [1]

8. (a) Write out and complete the sentences below about stem cells.

_____ stem cells can form the full range of cell types in the body, whereas _____ stem cells can only form cells of the same general type. [2]

(b) Meristems are the source of stem cells in plants. State precisely the location of meristems in plants. [1]

9. Apart from a possible need to carry out pre-treatment involving chemotherapy and/or radiotherapy, describe **two** risks with using stem cells as part of a medical treatment. [2]

10. The cube shown below has sides of length 1 cm, a surface area of 6 cm² and a volume of 1 cm³.

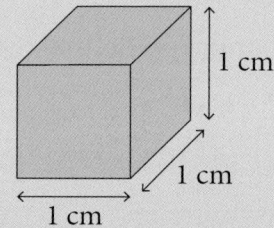

The table below shows how various cube dimensions change as the length of the sides increase.

Side length / cm	Area / cm²	Volume / cm³	Surface area to volume ratio
1	6	1	6
2	24	8	3
3		27	2
4	96		1.5
5	150	125	

(a) Draw and complete the table by calculating and adding the three missing values. [3]
(b) Describe the relationship between cube volume and its surface area to volume ratio. [1]
(c) Using the information in the table, explain why large multicellular animals have specialised surfaces (e.g. lungs in mammals) for the exchange of gases. [3]

1.2 Photosynthesis and Plants

1. Write out and complete the sentence below about photosynthesis by adding the missing words.

   During photosynthesis, plants produce sugars and starches using _____ energy. This is trapped by the _____ in chloroplasts. [2]

2. (a) Write the word equation for photosynthesis. [2]
   (b) Write the balanced chemical equation for photosynthesis. [2] **HT ONLY**

3. (a) The following statements describe stages in the starch test, but they are not in the correct order.

   A – dip the leaf in boiling water again
   B – place the leaf on a white tile and add iodine
   C – boil the leaf in ethanol
   D – add the leaf to boiling water for 30 seconds

   (i) Using the letters (**A**, **B**, **C** and **D**), state the order in which these stages would take place during the starch test. [1]
   (ii) Give the function of stage **C**. [1]
   (b) State the colour change that will occur during a positive starch test. [2]

4. (a) When carrying out photosynthesis investigations it is usually necessary to de-starch the plant first.
   (i) Describe how this is done. [1]
   (ii) Explain why it is necessary. [1]
   (b) (i) The partially-completed table below shows some of the stages when carrying out a starch test on a leaf. Copy out and complete the table.

Description of stage	Explanation of stage
place the leaf in boiling water initially	
	to remove the alcohol
dip the leaf in hot water	

   [3]
   (ii) Give **one** safety precaution which must be taken during the stages described in the table. [1]
   (iii) What further step(s) must be taken to show if a leaf contains starch? [2]

5. (a) Describe how you would set up an investigation to show that light is necessary for photosynthesis. (In your answer, you do not need to describe how to carry out a starch test). [4]
(b) (i) Describe what is meant by a variegated leaf. [1]
(ii) In what type of photosynthesis investigation would a variegated leaf be used? [1]
(iii) Describe the outcome you would expect in this investigation. [2]

6. The two sets of apparatus shown below can be used to compare the rates of photosynthesis in pondweed in different light intensities.

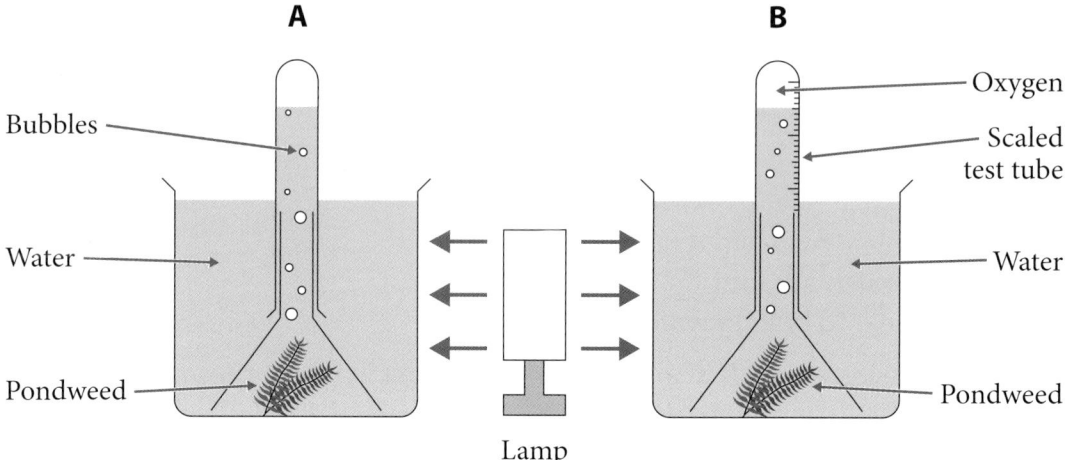

In setup **A** the number of oxygen bubbles given off can be used to compare the rates of photosynthesis in different light intensities. In **B** the volume of oxygen collected at the top of the test tube is measured using the scale.
(a) Describe **one** way in which the light intensity reaching the pondweed can be changed. [1]
(b) Suggest **two** reasons why the measurement of oxygen given off by the pondweed may be less accurate in setup **A**. [2]
(c) Give **two** variables which should be controlled when investigating the effect of light intensity on the rate of photosynthesis. [2]

7. The graph below shows the effect of light intensity and temperature on the rate of photosynthesis. This investigation was carried out at normal atmospheric carbon dioxide levels.

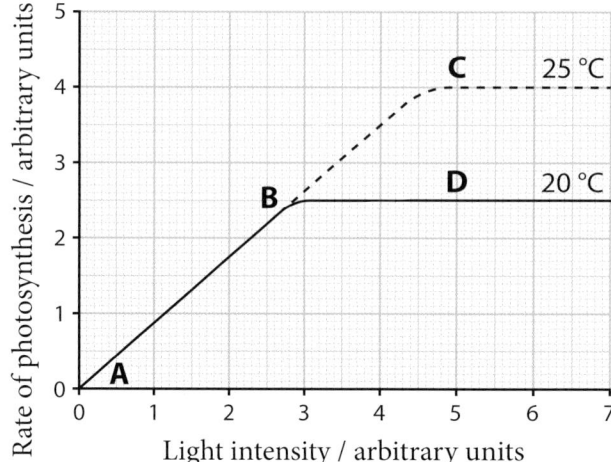

(a) Describe and explain the change in rate of photosynthesis between **A** and **B**. [2]
(b) (i) Calculate the percentage increase in rate of photosynthesis at **C** compared with **D**. [2]
    (ii) Explain this rate of increase in rate of photosynthesis at **C** compared with **D**. [2]
(c) Suggest **two** ways in which it would be possible to further increase the rate of photosynthesis at **C**. [2]

8. Hydrogencarbonate indicator is sensitive to changes in carbon dioxide level. The table below shows the colour of the indicator in different carbon dioxide levels.

Carbon dioxide level	Colour of hydrogencarbonate indicator
low	purple
normal atmospheric level	red
high	yellow

In an investigation, pondweed was placed in hydrogencarbonate indicator and the colour of the indicator monitored over a 24-hour period in a glasshouse. The results of the investigation are shown in the diagram below.

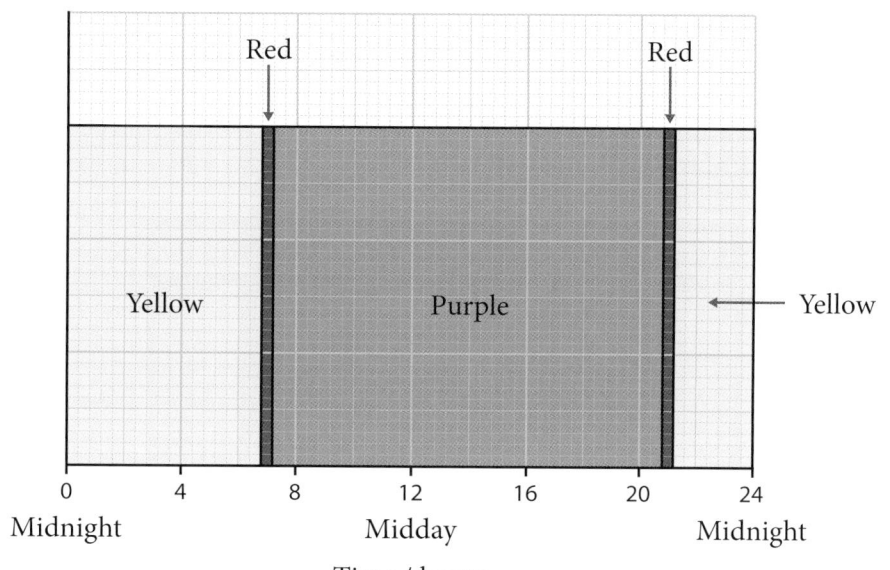

(a) For how long in the 24-hour period was the indicator purple? [1]
(b) Explain why the indicator was yellow during the times shown. [3]
(c) Explain why the indicator was purple during the time shown. [3]
(d) (i) Describe what is meant by the term compensation point. [2]
    (ii) On how many occasions was the pondweed at its compensation point in this 24-hour period? [1]

9. The diagram below represents a section through a leaf.

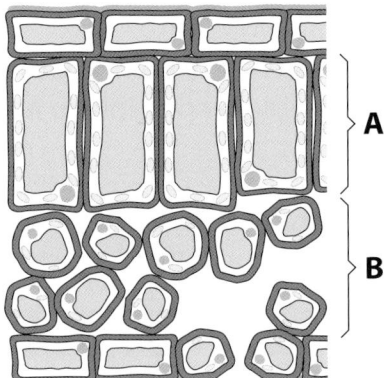

(a) Give **one** similarity and **one** difference in structure between the upper and lower epidermis. [2]
(b) (i) Name layers **A** and **B**. [2]
    (ii) Apart from being closer to the main source of light, give **two** reasons why the rate of photosynthesis is higher in layer **A** than layer **B**. [2]
(c) In many species the leaves are relatively long and wide giving a large surface area. Suggest **one** advantage and **one** disadvantage in leaves having a large surface area. [2]
(d) State **two** ways in which leaves are adapted for gas exchange. [2]

10. Leaves are structures highly adapted for photosynthesis. The diagram below represents a cross-section of a typical mesophytic leaf.

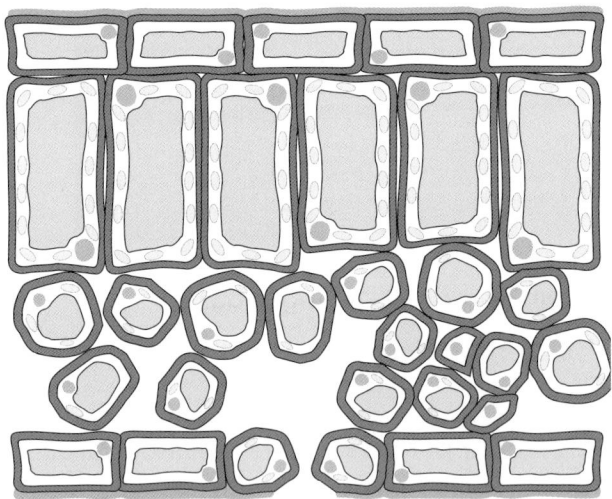

In terms of trapping light energy and in facilitating the diffusion of the gases involved in photosynthesis, using the diagram above, describe fully how leaves are adapted for photosynthesis.

In this question, you will be assessed on your written communication skills including the use of specialist scientific terms. [6]

1.3 Nutrition and Food Tests

1. (a) (i) Copy and complete the table below about food tests.

Test	Food type	Colour change if food type present
		Solution turns from blue to lilac/purple
	Starch	
Ethanol		

[6]

   (ii) Name the food test which requires heating. [1]

(b) Benedict's reagent turns from blue to brick red if sufficient reducing sugar is present. In an investigation, a student wished to compare the amount of reducing sugar in chocolate and in cake.

The student added equal masses of white chocolate and cake to Benedict's reagent in separate boiling tubes. The boiling tubes were then placed in the same beaker containing very hot water. At the end of the investigation the student compared the colour of the Benedict's reagent in the two boiling tubes.

   (i) Suggest why the student used white chocolate rather than dark brown chocolate. [1]

   (ii) Suggest why the boiling tubes containing each food type and the Benedict's reagent were placed in the same beaker. [1]

The first time the student carried out this investigation both boiling tubes turned brick red after heating. This showed that both foods contained large amounts of reducing sugar (not that they contained equal amounts).

   (iii) Suggest **two** ways in which the student could refine the investigation to find out which food contained more reducing sugar. [2]

2. (a) Glucose can be described as a 'simple' sugar. Suggest what is meant by 'simple' in this context. [1]

   (b) Lactose is an example of a disaccharide. Suggest what is meant by the term disaccharide. [1]

(c) The diagram below represents the energy storage molecule glycogen.

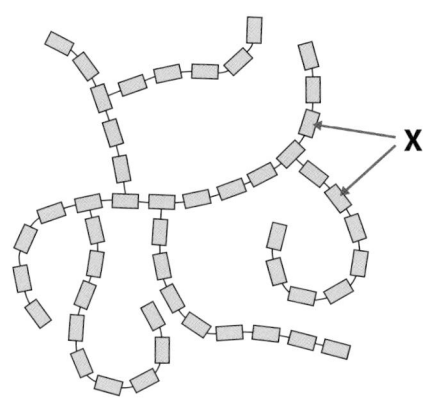

(i) State what the structures labelled **X** represent. [1]
(ii) Using the diagram, state **two** ways in which glycogen is adapted as a storage molecule. [2]
(iii) Name **one** other carbohydrate energy storage molecule. [1]

3. (a) Name the basic sub-unit of protein. [1]
   (b) Name **two** types of protein that have functional roles. [2]

4. (a) The diagram below represents a molecule of fat.

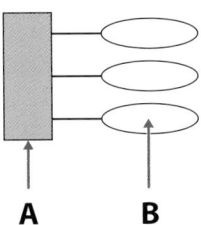

A   B

Identify the sub-units **A** and **B**. [2]

(b) The apparatus below was used to compare the energy content of two foods (**X** and **Y**). Both foods contained fat and protein, but one had a higher fat : protein ratio.

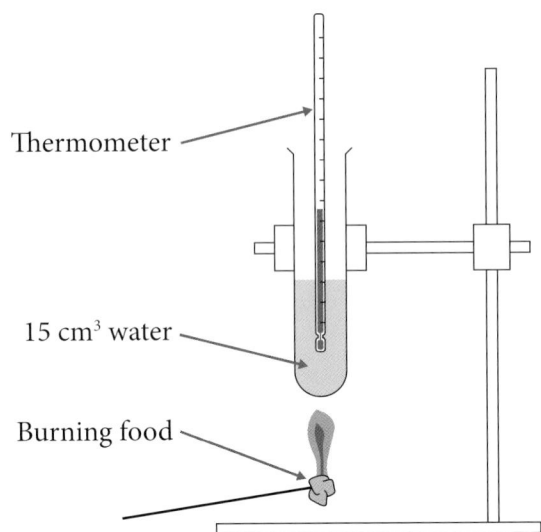

Thermometer

15 cm³ water

Burning food

A small portion of each type of food (**X** and **Y**) was burnt as shown and the results of the investigation are shown in the table below.

Food	Temperature of water / °C			
	Before burning	After burning	Increase	% increase
X	18	28	10	55.6
Y	19	35	16	

(i) Calculate the percentage increase in water temperature caused by burning food **Y**. [2]

(ii) The energy released by each food can be calculated using the formula:

energy (J) = volume of water (cm³) × temperature increase (°C) × 4.2

Calculate the amount of energy released during the burning of food **X**. [2]

(iii) Suggest which food had the higher fat : protein ratio. Explain your answer. [2]

(iv) State **two** variables which should have been controlled in this investigation. [2]

1.4 Enzymes and Digestion

1. Copy and complete the sentences below about enzymes.

   Enzymes are referred to as _____ as they speed up the rate of reactions without being used up in the reaction itself. Enzymes are formed of _____ molecules. [2]

2. The diagram below represents the action of an enzyme.

   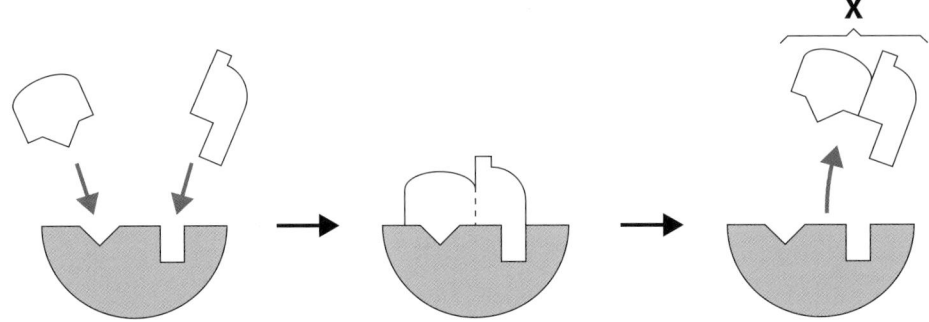

   (a) Identify the structure labelled **X**. [1]
   (b) What name is given to this model of enzyme action? [1]
   (c) Explain how this model of enzyme action can be used to explain enzyme specificity. [2]
   (d) The enzyme salivary amylase breaks down starch in food into its component sugars. Explain why the enzyme in the diagram is unlikely to be salivary amylase. [1]

3. The diagram below shows how enzyme concentration affects the rate of a reaction.

   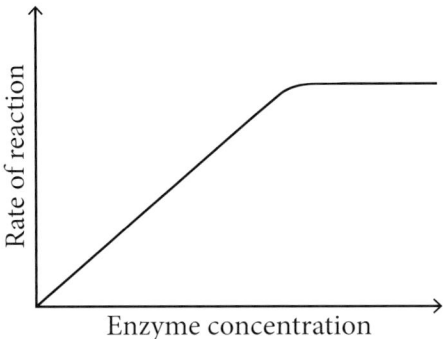

   (a) Describe and explain the results shown. [4]

(b) The second diagram shows the effect of adding a fixed amount of an inhibitor.

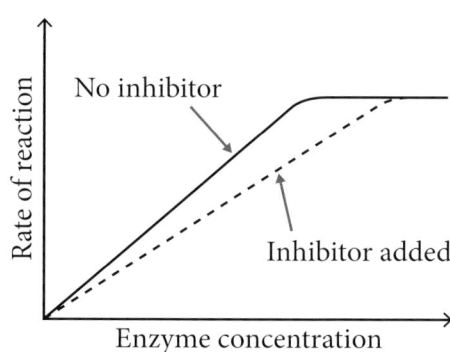

  (i) Describe and explain the effect of the inhibitor. [3]
  (ii) Suggest what would have happened if the amount of inhibitor added had been doubled. [2]

4. (a) The graph below shows the effect of temperature on enzyme activity.

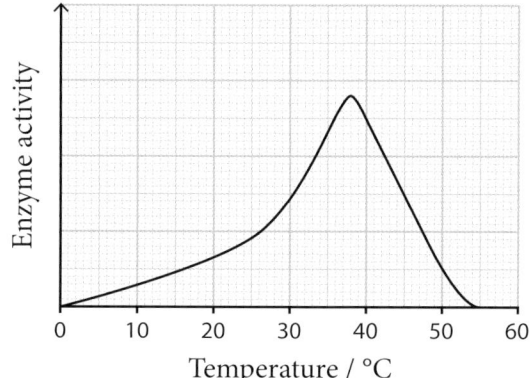

  (i) Identify the optimum temperature for this enzyme. [1]
  (ii) Describe and explain the difference in rate of reaction at 30 °C compared to 20 °C. [2]
  (iii) Explain why there is reduced enzyme activity at 45 °C. [2]

(b) Biological washing powders contain enzymes. These washing powders usually contain a range of enzymes including lipases.
  (i) Name the type of molecule which is broken down by lipase. [1]
  (ii) Name the breakdown products following lipase action. [1]
  (iii) Suggest **one** reason why some people prefer non-biological washing powders. [1]

5. In an investigation, the effectiveness of two protease enzymes (**A** and **B**) in breaking down protein was investigated at a range of temperatures. Equal volumes of each enzyme were added to equal volumes of protein in separate boiling tubes and the length of time taken to fully break down the protein was recorded. This was repeated for a range of temperatures and the results are shown in the table below.

Temperature / °C	Time to break down protein / minutes	
	Enzyme A	Enzyme B
20	27	33
30	19	22
40	12	17
50	36	46
60	53	No result

(a) In terms of the effect of temperature on enzyme activity, describe and explain the results for enzyme **B**. [4]

(b) Enzyme **A** is preferred for use in biological washing powders. Using the information in the table, suggest reasons why enzyme **A** is preferred to enzyme **B** in biological washing powders. [3]

6. (a) Describe the difference between the processes digestion and absorption. [2]

(b) The diagram below represents part of a villus.

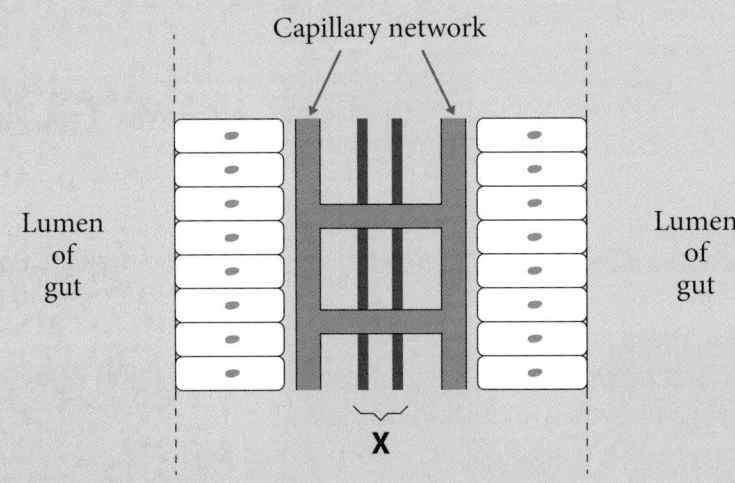

(i) Name structure **X**. [1]

(ii) Using the diagram only, state **three** ways in which villi are adapted for absorption. [3]

(iii) Name the part of the digestive system in which villi occur. [1]

1.5 The Respiratory System, Breathing and Respiration

1. State the word or words which best matches each of the following descriptions.
   - Surfaces in living organisms across which gas exchange takes place.
   - The gas produced in aerobic respiration.
   - The liquid produced in anaerobic respiration in yeast. [3]

2. **(a)** The diagram below represents part of the respiratory system following inhalation (breathing in).

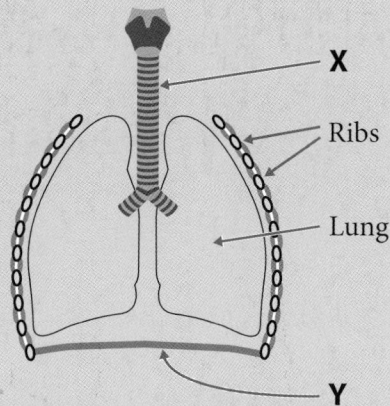

   (i) Name structures **X** and **Y**. [2]
   (ii) During exhalation (breathing out) the lungs will decrease in size. Using the diagram, state **two** other changes that will take place in the respiratory system when exhaling. [2]

   (b) The diagram below represents an alveolus and a blood capillary.

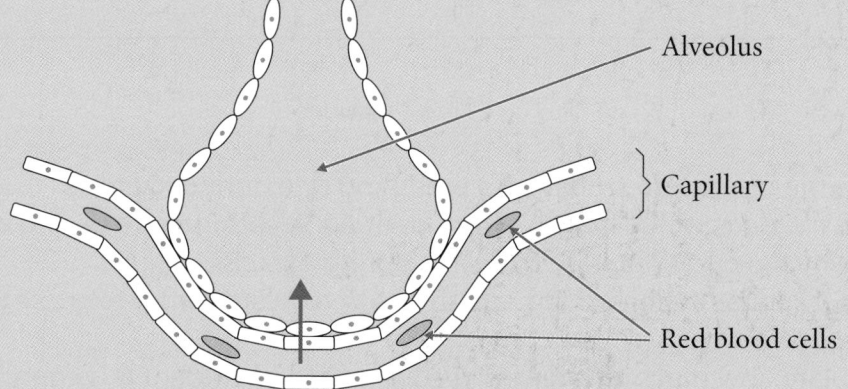

   (i) Name the gas which diffuses in the direction of the arrow. [1]
   (ii) Using your understanding of breathing, explain why this gas diffuses in this direction. [2]
   (iii) Using the diagram **only**, give **two** ways in which gas exchange is facilitated between the alveolus and the capillary. [2]

UNIT 1: CELLS, LIVING PROCESSES AND BIODIVERSITY

HT ONLY

3. (a) State the balanced chemical equation for aerobic respiration. [2]
    (b) (i) State the function of mitochondria in cells. [1]
        (ii) Suggest why there are many more mitochondria in 1 mm³ of muscle cell than in 1 mm³ of skin cell. [2]
        (iii) Suggest why 1 mm³ in the two cell types was compared, rather than just the total number of mitochondria per cell. [1]

4. The graph below shows the effect of exercise on the breathing rate of two students (**A** and **B**).

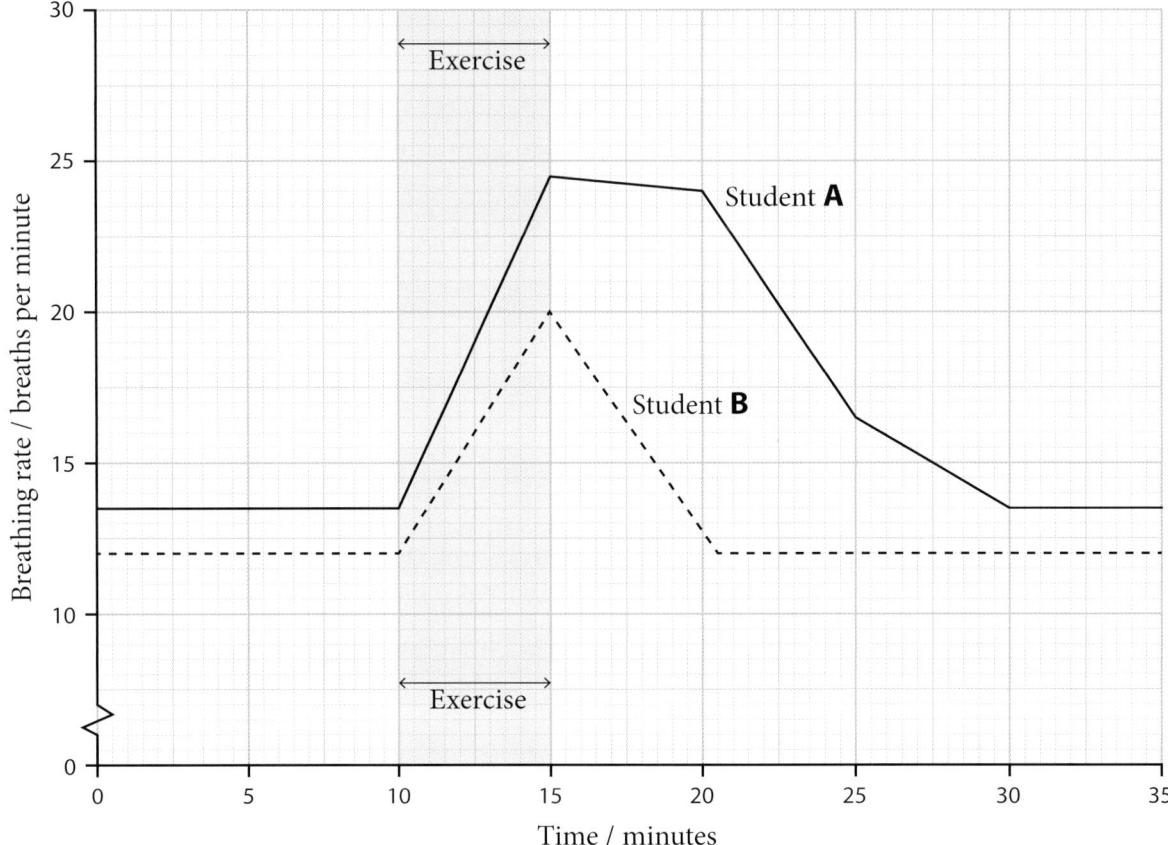

   (a) Calculate the increase in student **A**'s breathing rate during exercise. [2]
   (b) Calculate the percentage increase in student **A**'s breathing rate during exercise. [2]
   (c) Explain why the breathing rate increases during exercise. [3]
   (d) The time taken for the breathing rate to return to normal after exercise is described as the recovery time.
       (i) Calculate how much shorter the recovery time of student **B** is compared to that of student **A**. [2]
       (ii) It is thought that student **B**'s recovery time is shorter due to exercising more often and being fitter. Use the graph to provide **one** other piece of evidence which suggests that student **B** is fitter than student **A**. [1]
   (e) The effect of exercise on breathing **rate** was analysed in the investigation above. State **one** other effect of exercise on breathing. [1]

5. The diagram below shows apparatus that can be used to investigate respiration in yeast.

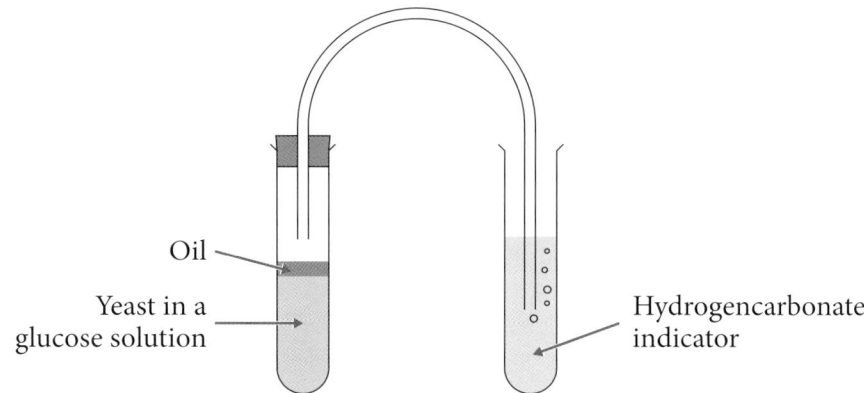

(a) Name the type of respiration being investigated. Explain your answer. [2]

(b) (i) Name **one** way in which the rate of respiration could be measured in this investigation. [1]

(ii) Describe how you could use this apparatus to compare the rates of respiration at 20 °C and 30 °C. Your answer should include **two** variables that you would need to control. [4]

1.6 Nervous System and Hormones

1. (a) The diagram below shows how some structures involved in nervous communication are linked.

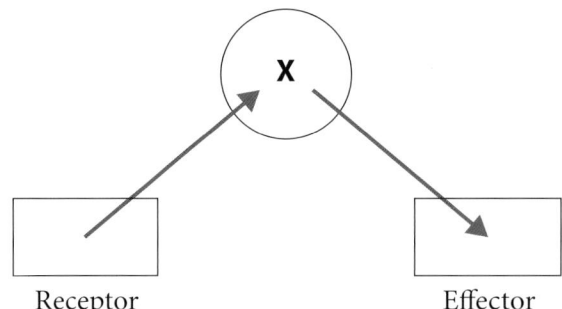

(i) Name the part of the nervous system represented by **X** in the diagram. [1]
(ii) Name **one** part of the body that could be **X**. [1]
(iii) Suggest what the arrows represent. [1]

(b) State **two** ways in which voluntary actions are normally different from reflex actions. [2]

2. (a) Draw and complete the table below by either naming the appropriate parts of the eye or describing their functions.

Part of eye	Function
cornea	
	contains the cells which are sensitive to different types of light
iris	
	transfers nerve impulses from the eye to the brain

[4]

(b) The aqueous humour is transparent.
(i) Explain fully why this is important. [2]
(ii) Give the function of the vitreous humour. [1]

HT ONLY

(c) The diagram below represents the ciliary muscles, sensory ligaments and lens when viewing a close-up (near) object.

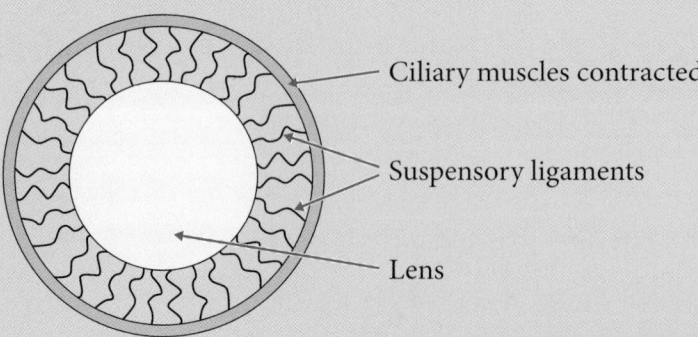

(i) Explain how these structures change when viewing a distant object. [3]
(ii) Explain how these changes allow the eye to focus on a distant object. [2]

3. The diagram below represents a neurone.

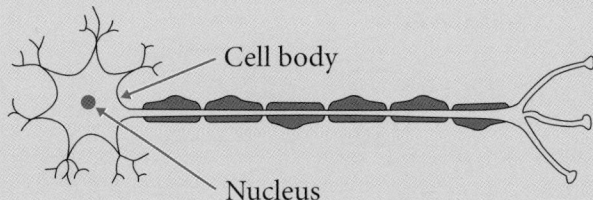

(a) State the function of a neurone. [1]
(b) The neurone in the diagram has a cell body with many branched ends. Give the function of the branched ends. [1]
(c) Apart from having a cell body and many branched ends, give **two** ways in which the neurone in the diagram is adapted for its function. [2]

4. The diagram below represents a synapse between two neurones. Small vesicles produce and release transmitter chemical into the gap between the neurones.

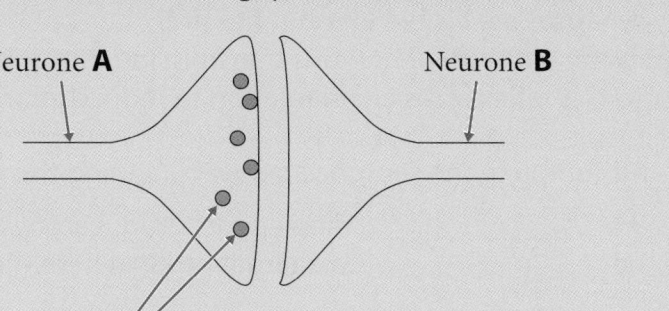

(a) Describe and explain the role of the transmitter chemical in nervous transmission. [3]
(b) What is the evidence that the direction of nervous transmission is from neurone **A** to neurone **B**? Explain your answer. [2]
(c) A particular medical condition causes the gap between neurones to be slightly wider than normal. Suggest the effect that this will have on nervous transmission. Explain your answer. [2]

UNIT 1: CELLS, LIVING PROCESSES AND BIODIVERSITY

5. The diagram below represents the spinal reflex arc.

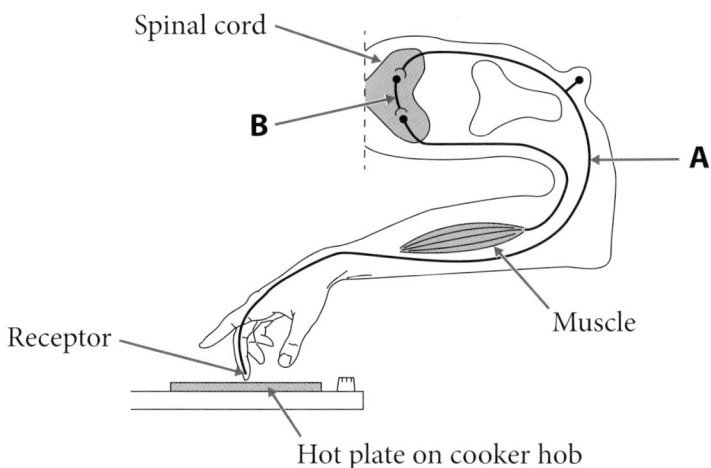

(a) Name the neurones labelled **A** and **B**. [2]
(b) How many synapses are visible in this reflex arc? [1]
(c) Describe fully the pathway of nervous transmission through this reflex arc. [5]
(d) Describe how this reflex arc is an adaptation to reduce injury. [2]

6. (a) Write out and complete the sentence below by adding the missing words.

Hormones are _____ messengers produced by glands which are carried by the _____ to a target organ where they act. [2]

(b) Insulin is an example of a hormone.
  (i)   Name the organ in the body which produces insulin. [1]
  (ii) What causes this organ to increase insulin production? [1]
  (iii) State **two** ways in which insulin reduces blood sugar levels. [2]

7. Diabetes is a condition in which the blood glucose control mechanism fails. The graph opposite shows how blood glucose levels change in an individual who has diabetes and in an individual who does not have diabetes.

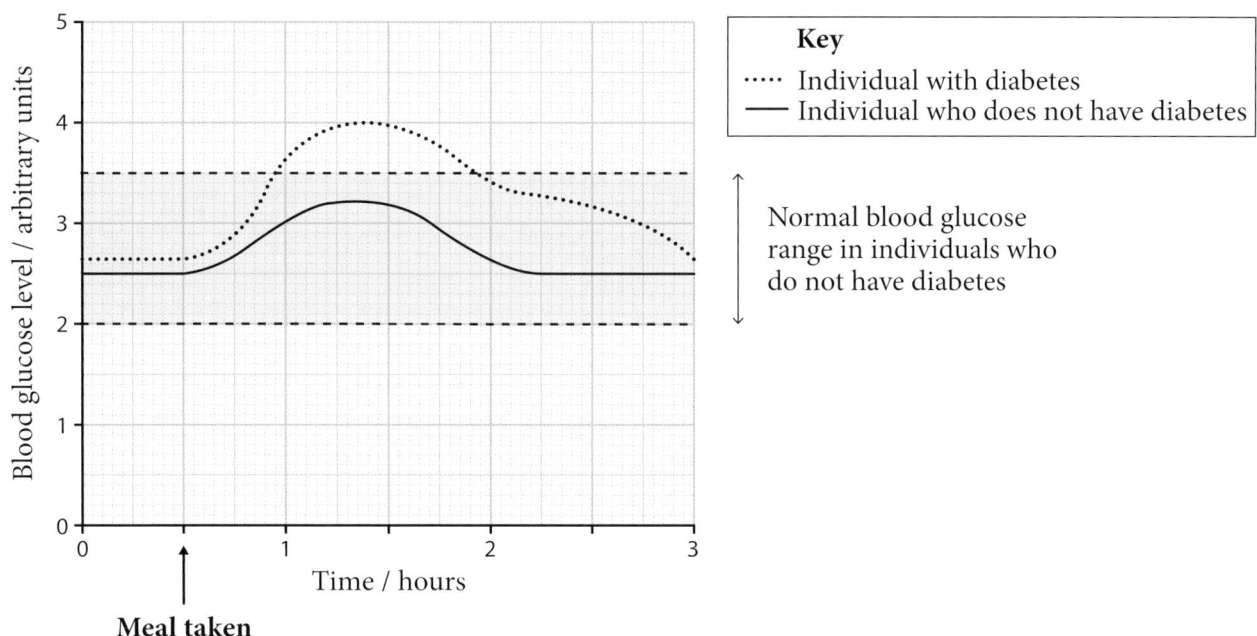

Meal taken

(a) Summarise the results shown. [4]

(b) There are two distinct types of diabetes, Type 1 and Type 2.
  (i) Which type is linked to lifestyle factors? [1]
  (ii) Describe the difference in treatment of the two conditions. [2]

The table below shows the number of people being treated for diabetes in a council area over a 10-year period.

Year	Number of people being treated for diabetes	
	Type 1	Type 2
2010	30	121
2012	31	139
2014	34	168
2016	38	194
2018	41	220
2020	44	245

  (iii) Calculate the percentage increase in the number of people being treated for diabetes in this council area over the 10-year period. Give your answer to three significant figures. [3]
  (iv) Give **three** conclusions that can be made from the data in the table. [3]

(c) Diabetes is a condition that costs the health service very large amounts of money. Apart from reference to the number of people who have the condition, suggest **two** other reasons for the very high cost. [2]

UNIT 1: CELLS, LIVING PROCESSES AND BIODIVERSITY

8. (a) The diagram below represents the excretory system.

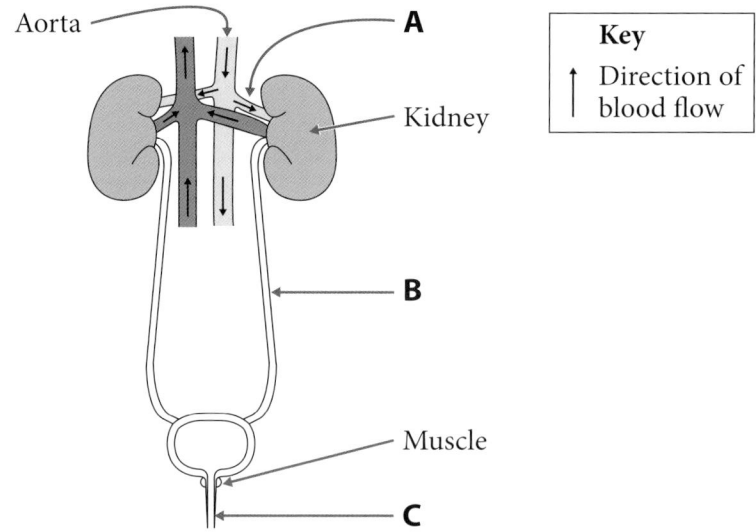

    (i) Name the structures labelled **A**, **B** and **C**. [3]
    (ii) Suggest the function of the muscle shown on the diagram. [2]
 (b) (i) Give the **two** main functions of the kidney. [2]
    (ii) The body loses water through urine release. State **two** other ways in which the body loses water. [2]

9. The antidiuretic hormone (ADH) is involved in osmoregulation in the kidney.
 (a) (i) Describe the relationship between ADH concentration in the blood and volume of urine production. [1]
    (ii) Explain this relationship. [1]
 (b) The diagram opposite shows how the ADH concentration in an individual's blood changed after a period of intense exercise.
    (i) For how long did the ADH concentration in the blood remain raised following exercise? [1]
    (ii) In terms of the body's water balance, explain the reason for the changes in ADH concentration shown. [4]

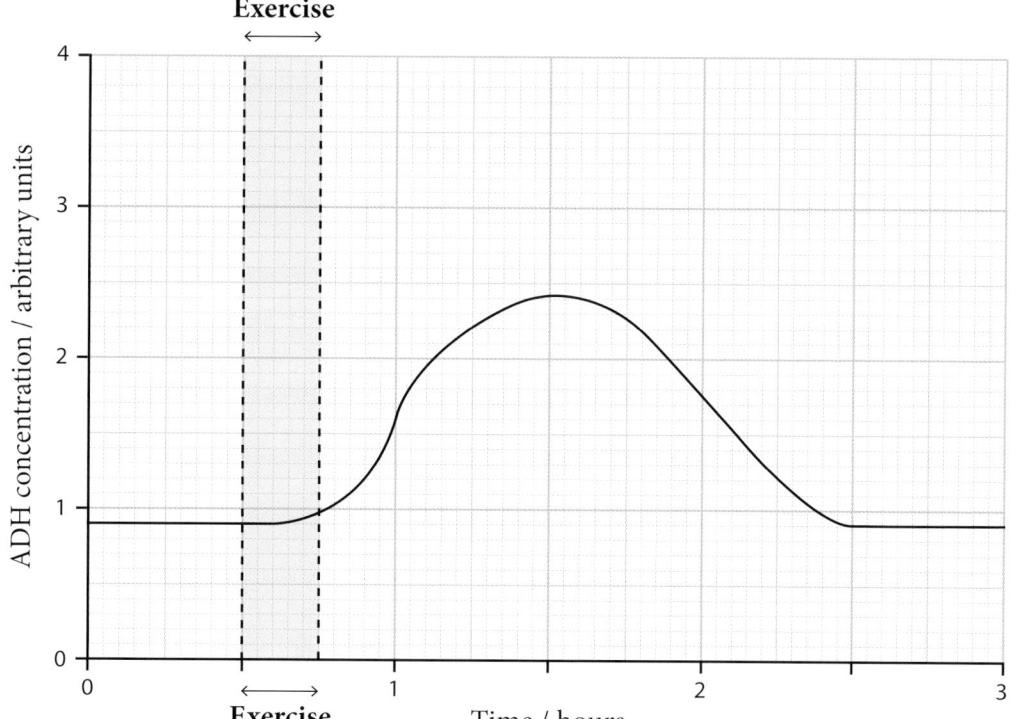

10. Plants receiving unidirectional light (from one side only) grow in the direction of the light source.
    (a) What name is given to this response? [1]
    (b) Explain the advantage of this response. [2]
    (c) Explain what causes this growth effect. [2]

UNIT 1: CELLS, LIVING PROCESSES AND BIODIVERSITY

11. An investigation was set up as shown below to study the effect of unidirectional light on plant seedlings. In both setup **A** and **B** seedlings were subjected to unidirectional light. However, in setup **B** the container holding the seedlings rotated at a very slow rate during the investigation.

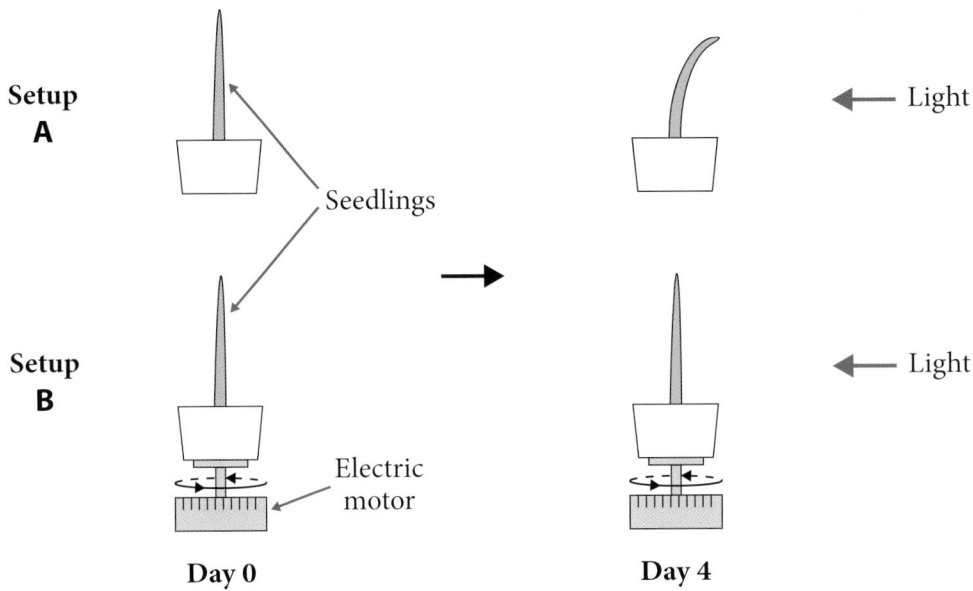

(a) Give **two** variables which should have been controlled in this investigation. [2]
(b) In terms of auxin movement and action, explain the results for setup **A**. [4]
(c) Explain the results for setup **B**. [2]

1.7 Ecological Relationships and Energy Flow

1. Copy and complete the table below about ecological terms and their descriptions.

Term	Description
Community	
	The number of organisms of a species in an area
Environment	
	The combination of all the organisms and their surroundings in an area

   [4]

2. (a) Soil pH is an abiotic factor.
   - (i) Define the term abiotic factor. [1]
   - (ii) Describe **one** way in which it is possible to measure the pH of soil. [1]

   (b) Lousewort is a plant found on moorland. It is a plant which is seldom more than a few centimetres in height.

   Lousewort surrounded by grass

   Moorlands have soils with a low pH (acidic soils). Conditions in moorland are often harsh with low temperatures, strong winds and high rainfall.

(i) Name the apparatus used to measure wind speed. [1]
(ii) Using the information provided, suggest **one** way in which Lousewort is adapted for its habitat. Explain your answer. [2]
(iii) Describe how you could sample the abundance of Lousewort in an area of moorland. [4]

3. In an investigation, five groups of students each used a belt transect to study how the number of plant species changed from the centre of a grass field to the centre of a small wood at the edge of the field.
   (a) Describe what is meant by the term belt transect. [1]
   (b) In the investigation, square quadrats of sides 50 cm were used. Calculate the area within each quadrat. Give your answer in square metres. [2]

The results of the investigation are shown in the graph below.

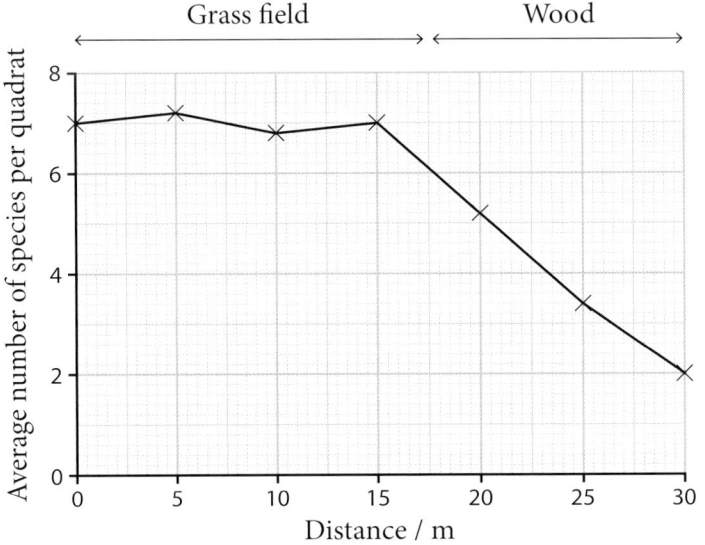

(c) Summarise the results shown. [3]
(d) It was proposed that light intensity was the main environmental factor affecting the number of plant species present. Suggest how the data supports this hypothesis. [2]

4. In an investigation of competition in plants, 100 very young seedlings of a plant species were planted in compost in a seed tray. The number of seedlings surviving, and their average masses, were measured and recorded over the following three years. The results are shown in the table below.

Year	Number of seedlings	Average mass of each seedling / g
1	100	0.6
2	61	2.8
3	27	7.4
4	14	16.9

(a) Describe the trends shown by the results. [2]
(b) Explain the results for number of seedlings over time. [3]

5. The flow diagram below represents four organisms in a food chain.

Hosta plant → slug → hedgehog → fox

(a) Define the term producer. [2]
(b) Name the organism which is a primary consumer. [1]
(c) Name the organism which feeds at trophic level 2. [1]
(d) State the initial source of energy in food chains. [1]

6. (a) The diagram below represents energy flow through part of a woodland ecosystem. Values are in kJ.

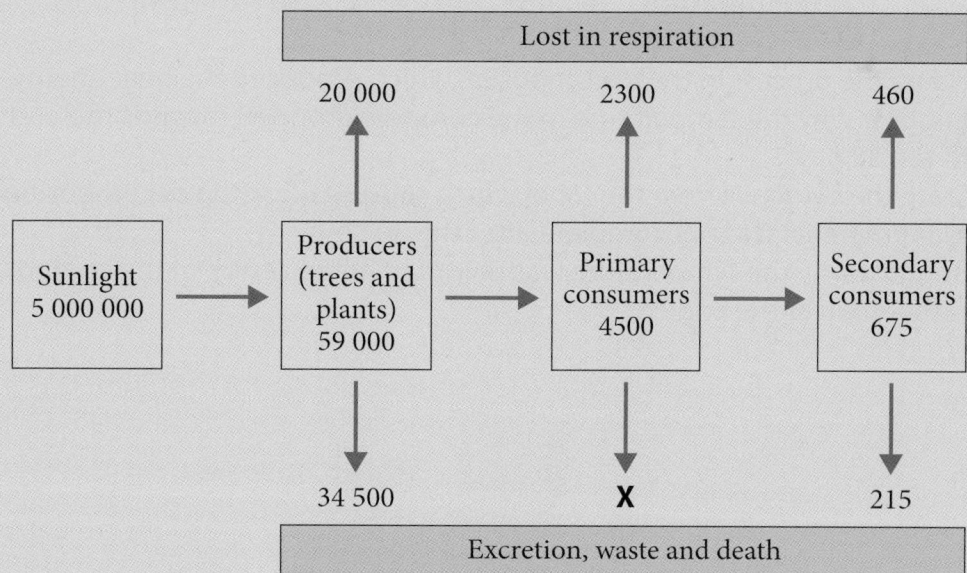

(i) Suggest **one** reason why only a small proportion of the energy in sunlight is available for photosynthesis by plant and tree leaves. [1]
(ii) Between which two trophic levels is there the greatest energy loss? [1]
(iii) Suggest **one** reason for this. [1]

(iv) Calculate value **X**. [2]
(v) Secondary consumers lose proportionally more of their energy as respiration than primary consumers. Suggest a reason for this. Explain your answer. [2]

(b) Transfer of energy between organisms can be represented by pyramids of numbers and pyramids of biomass. The diagram below represents a pyramid of numbers for the woodland ecosystem described above.

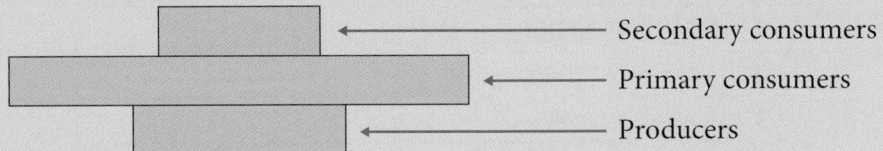

(i) Describe and explain how a pyramid of biomass would be different from the pyramid of numbers. [2]

HT ONLY
(ii) Explain why a pyramid of biomass may be a better representation in this ecosystem. [1]

7. (a) Saprophytic fungi are important in decomposing wood.
   (i) Describe the process of decomposition by these fungi. [3]
   (ii) Suggest why the rates of decomposition are faster in summer than in winter. [2]

(b) The table below shows the temperature changes that have taken place in the centre of a compost heap of grass cuttings and tree leaves over a period of 30 days. Bacteria and fungi were involved in decomposing the material in the compost.

Day	1	5	10	15	20	25	30
Temperature / °C	15	18	24	33	30	26	24

(i) Name the process in the decomposers which produced the heat energy. [1]
(ii) Suggest why the temperature in the compost dropped towards the end of the 30 days. [1]
(iii) State the evidence from the table which suggests that the decomposition of the plant material was not complete after the 30 days. [1]
(iv) Suggest why the temperature was recorded at the centre of the compost heap rather than at the edge. [1]

1.7 ECOLOGICAL RELATIONSHIPS AND ENERGY FLOW

8. The diagram below shows a simplified version of the carbon cycle.

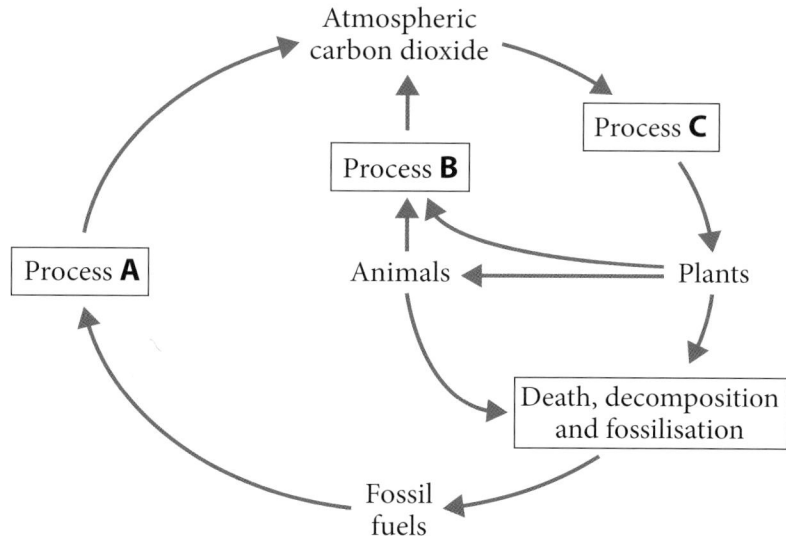

(a) Name processes **A**, **B** and **C**. [3]
(b) Name **two** types of decomposers. [2]
(c) Using the diagram, state **two** ways in which atmospheric carbon dioxide levels can be reduced. [2]

9. Willow is fast-growing tree which is commonly used as a biofuel. The stems of the willow can be cut into small sections or chips which can be used as a fuel in heating systems.

During harvesting, each tree is cut just above ground level allowing the next willow crop to grow rapidly from the cut tree stump. This means that each crop can be harvested after three or four years.

The use of willow as a biofuel is encouraged as it is regarded as a 'carbon neutral' fuel in that it doesn't contribute to the rise in carbon dioxide levels in the atmosphere when its whole life cycle is taken into consideration.

(a) Living trees both take in and give out carbon dioxide.
  (i) Name the process that takes in carbon dioxide. [1]
  (ii) Name the process that gives out carbon dioxide. [1]

(b) The graph below shows the overall balance of carbon dioxide intake and output in a four-year growth cycle in willow. Net intake is where there is an overall intake of carbon dioxide by the willow during that year. Net output is where there is an overall production of carbon dioxide that year (i.e. there is more carbon dioxide given out than is taken in by each willow tree).

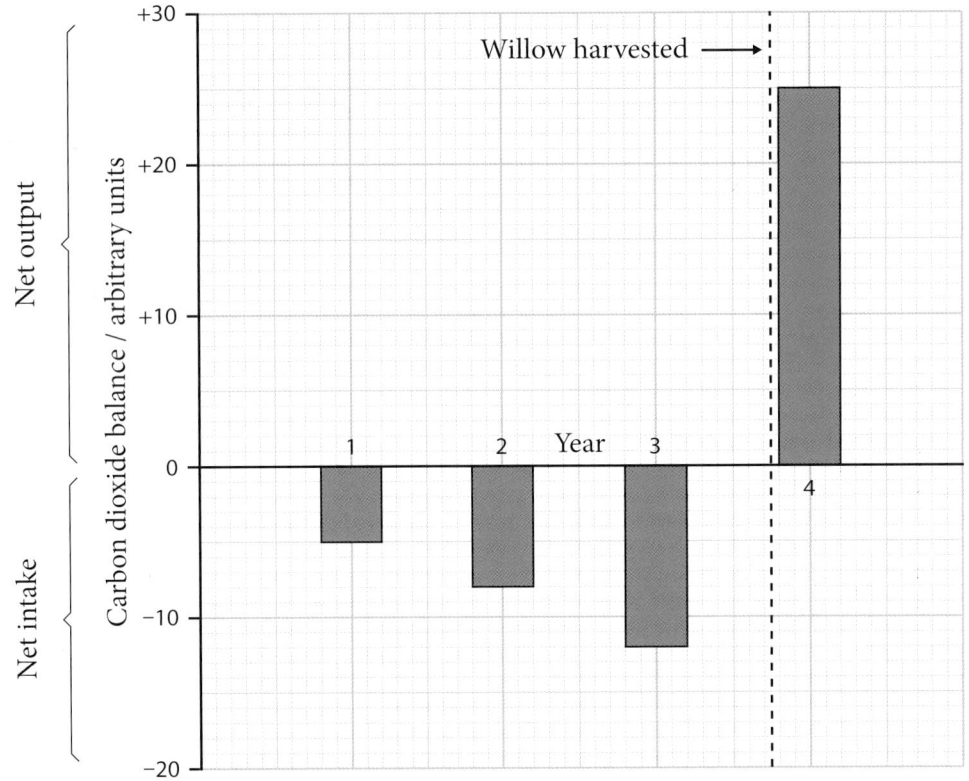

(i) Suggest why the net intake of carbon dioxide by each willow plant increases from year 1 to year 3. Explain your answer. [2]

(ii) Name the process that mainly contributes to the net output of carbon dioxide in year 4. [1]

(iii) Using the information provided in the graph, explain why using willow as a biofuel can be described as being 'carbon-neutral'. Explain your answer. [2]

HT ONLY **10.** Give an account of global warming. Your answer should describe its causes and the problems it produces.

In this question, you will be assessed on your written communication skills including the use of specialist scientific terms. [6]

11. (a) The diagram below shows a simplified nitrogen cycle.

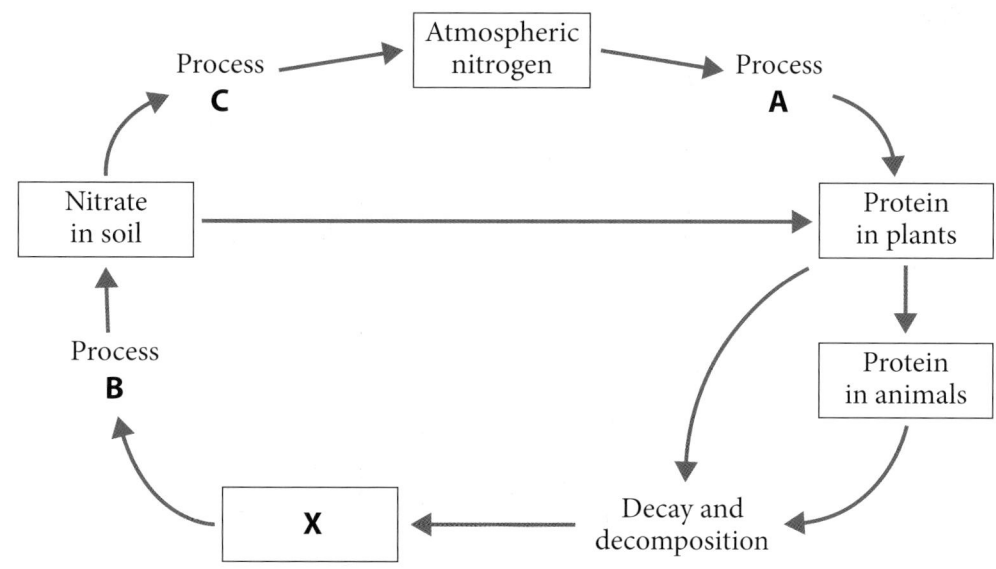

    (i) Identify the substance(s) labelled **X**. [1]
    (ii) Identify processes **A**, **B** and **C**. [3]

(b) The graph below shows how the numbers of two types of bacteria involved in the nitrogen cycle change as soil moisture content changes.

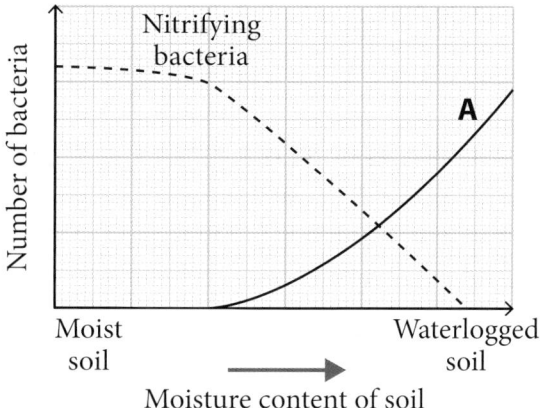

    (i) Identify the type of bacteria labelled **A**. [1]
    (ii) Describe and explain the distribution of bacteria **A**. [3]
    (iii) Describe the function of nitrifying bacteria and explain their change in number in the graph. [3]
    (iv) Waterlogged soils typically have low fertility. Using the graph, explain this statement. [4]

12. There are many examples of bog habitats in Ireland. Bogs are very wet and have peat rather than soil as the base on which plants grow. Bog habitats have very low fertility and very low levels of nitrate.

  (a) Using the information provided and your knowledge of the microorganisms involved in the nitrogen cycle, explain fully why in bog habitats nitrate levels are very low. [4]

  (b) The photograph below shows a Sundew plant.

Sundews are only found in bog habitats. They trap insects in their spoon-shaped leaves which contain numerous hairs that secrete a sticky glue-like substance. Once trapped the protein-rich insect is then broken down by enzymes released from the leaf.

  (i) Using the information provided, suggest why Sundews can survive in infertile bog habitats. [4]

  (ii) Suggest why Sundews are not found in other habitats such as grassland. [1]

13. Plants need to be able to absorb a range of minerals from the soil to grow properly.

  (a) Copy and complete the table below about minerals required by plants and their functions.

Mineral	Function
magnesium	
	required for making plant cell walls

[2]

  (b) Soil fertility can be increased by adding minerals to the soil in the form of fertiliser. Fertilisers can be natural, such as farmyard manure, or artificial.
  Give **two** advantages in using artificial fertiliser and **one** disadvantage. [3]

14. Root hair cells are specialised cells for the absorption of water and minerals. They are found on the surface of plant roots.
  (a) (i) Describe how these cells are adapted for their function. [1]
  (ii) Explain how this adaptation helps increase water and mineral uptake. [1]

  (b) The concentration of each mineral required for growth is normally higher in root hair cells than in the surrounding soil. Due to this concentration gradient, minerals are unable to enter root hair cells by diffusion.
  (i) Name the process by which root hair cells transport minerals from the soil into the root. [1]

  Respiratory poisons are substances which prevent respiration taking place in cells.
  (ii) Using the information provided and your knowledge, explain why root hair cells treated with respiratory poisons are unable to transport minerals from the soil into the root. [2]

15. The graph below shows the effect of a sewage leak entering a river on the river's oxygen levels.

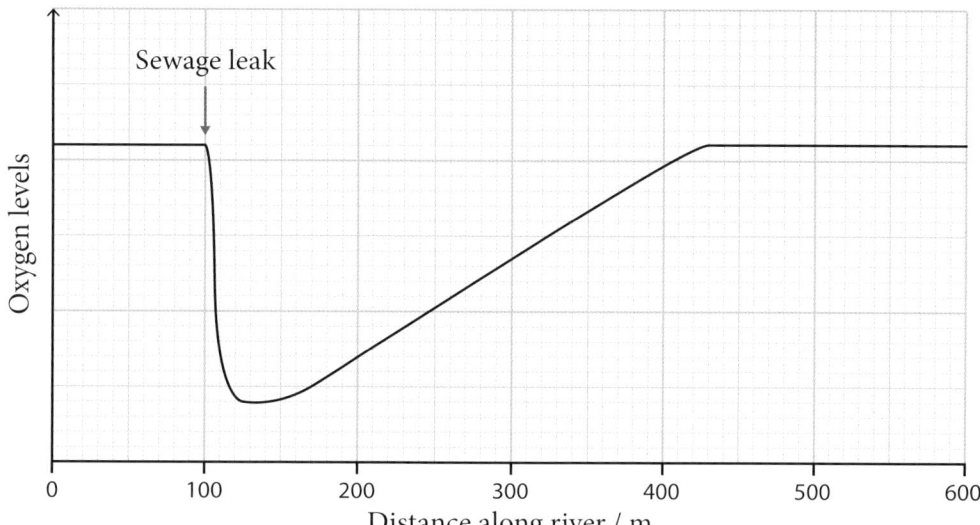

  (a) Calculate the length of river that was affected by sewage. [2]
  (b) Explain what causes the reduction in oxygen levels after the sewage entered the water. [2]
  (c) Suggest **one** reason why oxygen levels eventually returned to normal downstream. [1]

16. Fertiliser run-off entering rivers and lakes can have a harmful effect on biodiversity.
  (a) Nitrates in the fertiliser can stimulate the growth of aquatic plants and algae What name is given to this process? [1]
  (b) As a result of nitrate depletion over time and shading by other plants growing on the water surface, many of the aquatic plants and algae die. Describe and explain the sequence of events which typically follows the death of the plants and algae. [3]
  (c) Suggest **one** way in which fertiliser run-off into waterways can be reduced or stopped. [1]

Unit 2
Body Systems, Genetics, Microorganisms and Health

2.1 Osmosis and Plant Transport

1. **(a)** Define the term osmosis. [2]
   (b) The graph below shows changes in the average length of plant cells in a section of epidermal tissue after it had been placed in pure water.

   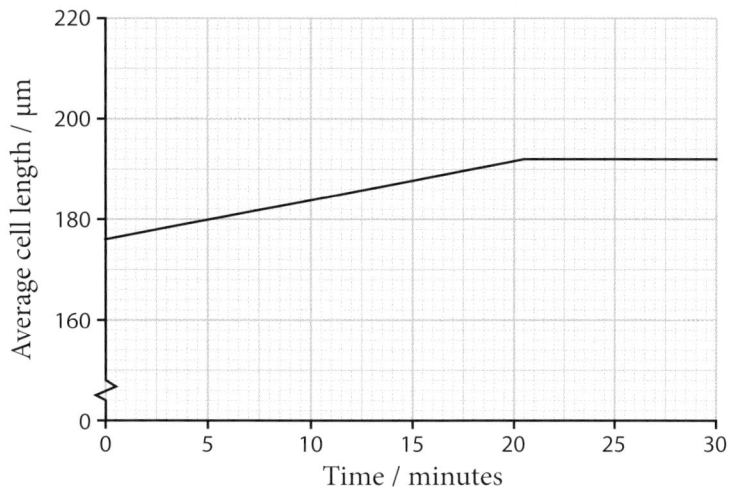

   (i) After how many minutes was full turgor achieved? [1]
   (ii) Calculate the percentage increase in average cell length during the investigation. Give your answer to one decimal place. [2]
   (iii) In terms of osmosis, explain the change in cell length after the epidermal tissue was placed in water. [3]
   (iv) Explain why the average length of epidermal cells levelled off. [1]
   (c) (i) Predict the results that would have been expected had the epidermal cells been placed in a strong sugar solution rather than in pure water. [2]
   (ii) Explain your answer to part (i). [3]

2. Five potato cylinders were cut to approximately 5 cm in length. Each cylinder was then placed in either water or a particular concentration of sucrose in a beaker as shown in the table below. After one hour, the cylinders were removed from the beakers and their lengths measured. The percentage change in length of each cylinder is recorded in the table below.

Solution / % sucrose	Change in potato length / %
0 (water)	+ 5.2
3	+ 2.8
6	+ 1.2
9	– 1.4
12	– 3.7

   (a) Explain the results for the potato cylinders which increased in length during the investigation. [3]

(b) Which concentration of sucrose used is closest to the concentration of the potato cells? Explain your answer. [3]

(c) Suggest **two** reasons why it may be more accurate to measure the mass of the potato cylinders rather than their length. [2]

3. A student is given two different concentrations of sugar in separate Visking tubing (solutions **A** and **B**). One of the solutions is 2.5 % sugar and the other is 7.5 % sugar. However, the student does not know which concentration is in each Visking tubing. The student is also provided with a beaker containing a 10 % sugar solution.

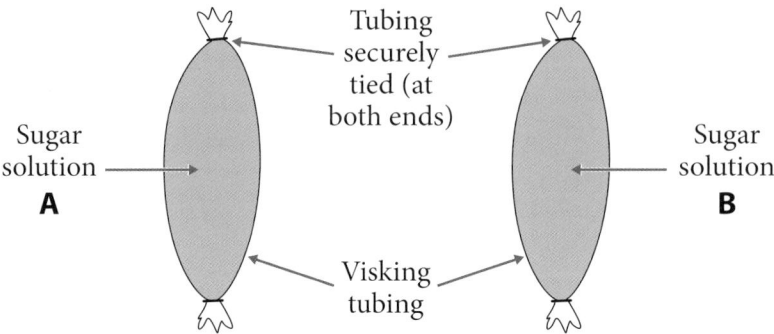

Using the materials supplied and any additional equipment required, describe and explain how the student could determine which of the Visking tubes contains the 2.5 % sugar. [4]

4. (a) Define the term transpiration. [2]

(b) Plant leaves are adapted to prevent excessively high rates of transpiration. Suggest **two** ways in which they are adapted to reduce transpiration rate. [2]

(c) In an investigation to compare the rates of water loss at different temperatures, four leaves were numbered, weighed, and then attached to a line of string by paper clips as shown below.

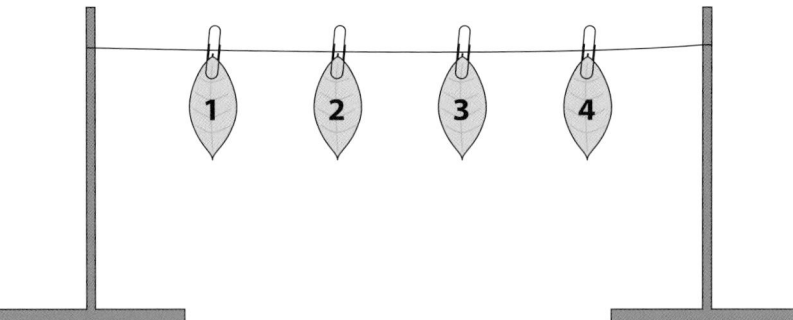

The leaves were kept at 15 °C for 24 hours and then re-weighed. At the same time, the process was repeated using four different leaves. These leaves were kept at 25 °C for 24 hours.

(i) At the start of the investigation the cut stalks of the leaves were smeared with petroleum jelly. Explain fully why this was done. [2]

(ii) Predict the results that you would expect in this investigation. Explain your answer. [2]

(iii) Apart from re-weighing both sets of leaves after 24 hours, give **two** variables which would need to be controlled in this investigation to ensure that the results were valid. [2]

5. The rate of transpiration in plants is affected by several environmental and other factors.
   (a) In an investigation, the transpiration rates of eight leaves of different surface area were measured. The results are shown in the table below.

Leaf surface area / mm²	Rate of transpiration / arbitrary units
550	18
673	25
402	10
499	14
620	20
523	16
683	27
560	19

   (i) Using the information provided, describe the effect of leaf surface area on transpiration rate. [1]
   (ii) Explain the results shown. [2]

   (b) An investigation was carried out to study the effect of wind speed on transpiration rate in a plant species. The results are shown in the graph below.

   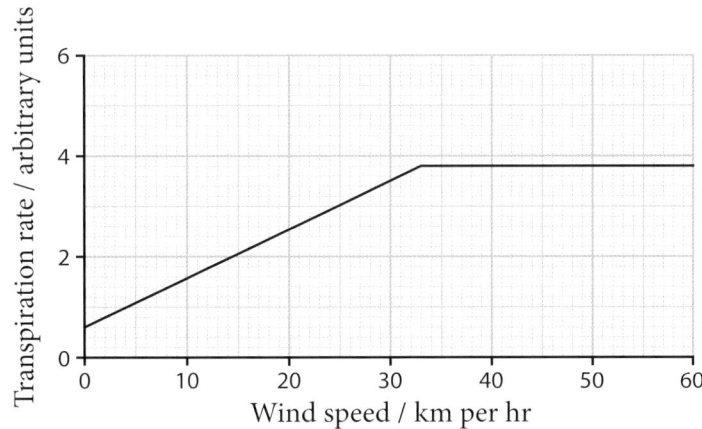

   (i) Using data, describe and explain the results shown. [3]
   (ii) Suggest the effect on the results shown if the investigation was carried out in conditions of higher humidity. [1]
   (iii) Apart from wind and humidity, name **one** other **environmental** factor which affects the rate of transpiration. [1]

2.2 The Circulatory System

1. Blood contains several different components, each of which has a different function.
   (a) The diagram below represents a red blood cell.

   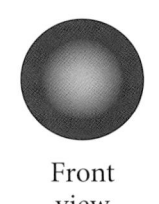

   Side view   Front view

   (i) Using **only** the diagram, give **one** way in which red blood cells are adapted for their function. Explain your answer. [2]
   (ii) Give **one** other way red blood cells are adapted for their function. Explain your answer. [2]
   (b) Copy and complete the sentence below about the role of platelets.

   Platelets help convert _____ to _____ at the site of a cut or wound. This conversion helps the blood to _____ and form a scab. [3]

   (c) Plasma transports many substances around the body. Name **two** of these substances. [2]

HT ONLY 2. It is important that the concentration of the blood is kept within a very narrow range. Describe and explain the effect on red blood cells should the blood become too dilute. [3]

3. (a) The diagram below represents a cross-section through an artery.

   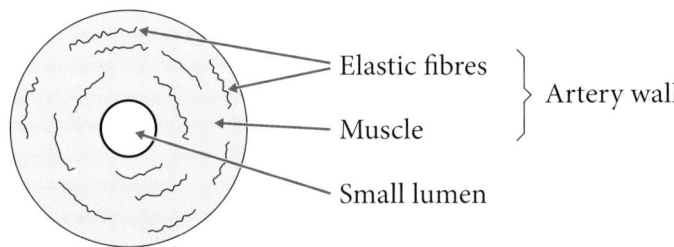

   (i) Describe and explain the role of the elastic fibres in the cell wall. [3]
   (ii) Suggest the benefit of there being a small (narrow) lumen. [1]
   (b) Veins, unlike arteries, have valves.
   (i) Describe the function of valves in veins and explain why they are needed. [2]
   (ii) Apart from the presence of valves, describe **two** other ways in which the structure of veins is different to the structure of arteries. [2]
   (c) Describe the function of capillaries and give **one** way in which they are adapted for this function. [2]

4. The diagram below shows how blood flows into, through, and out of the heart.

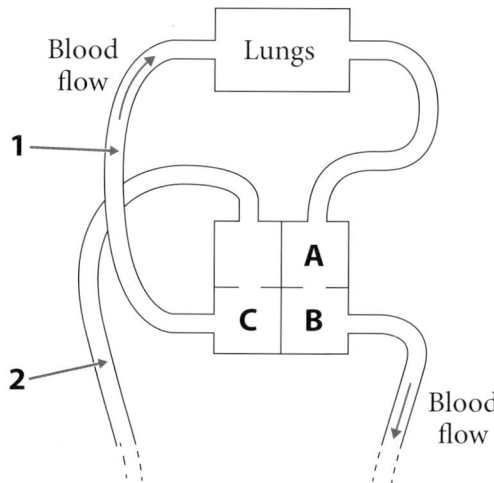

(a) Name the heart chambers **A** and **C**. [2]
(b) Although not visible in the diagram, the muscular walls of each heart chamber are not of the same thickness.
  (i) Describe and explain the difference in wall thickness of chambers **A** and **B**. [2]
  (ii) Describe and explain the difference in wall thickness of chambers **B** and **C**. [2]
(c) Name blood vessels **1** and **2**. [2]
(d) Name the blood vessel shown in the diagram which carries blood at the highest pressure. [1]
(e) Name the blood vessels which supply blood to the heart muscle. [1]

5. The graph below shows the pulse rates of three students before, during, and after they exercised for 10 minutes.

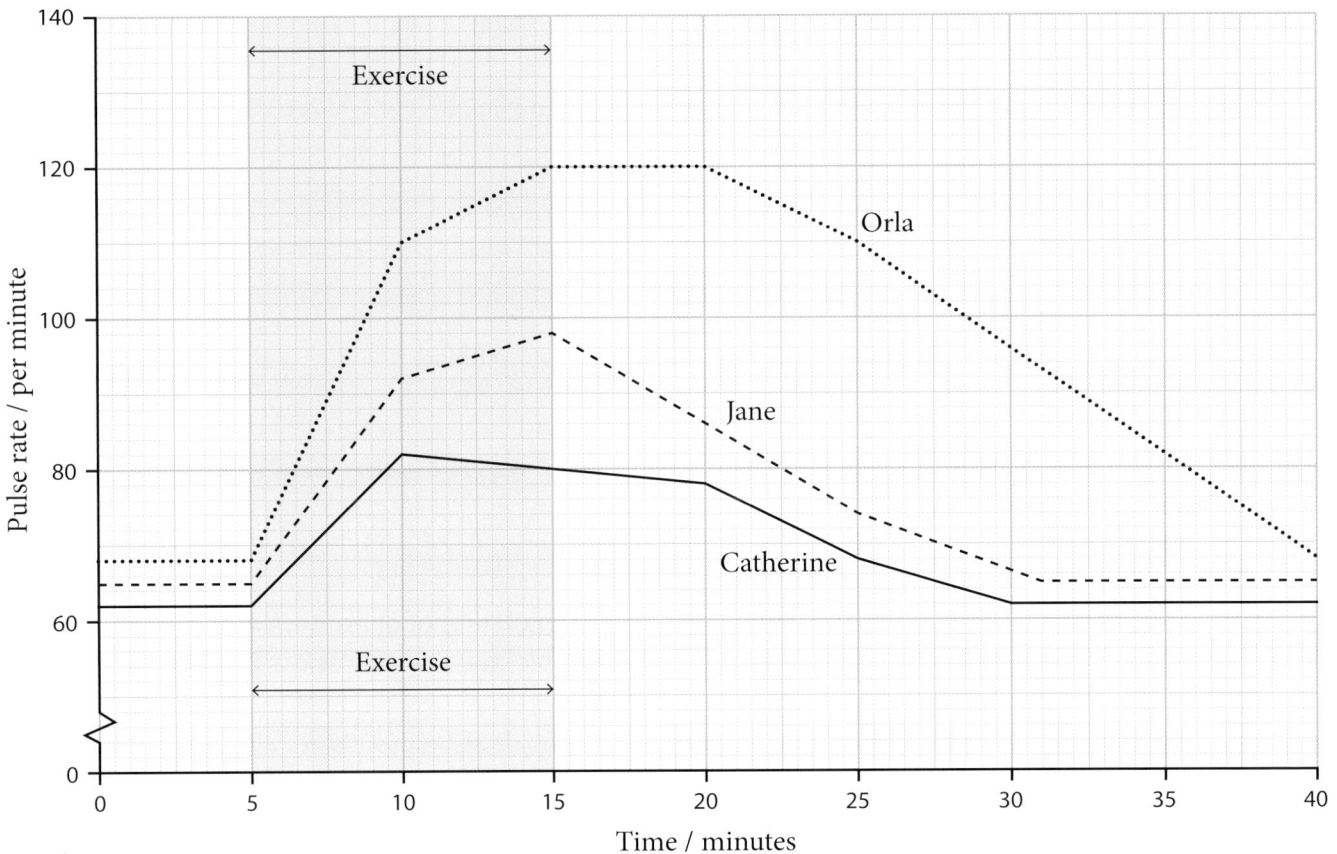

(a) Calculate the maximum increase in Catherine's pulse rate increase during exercise. [2]
(b) How long did it take her pulse rate to return to normal after exercise? [1]
(c) Catherine and Jane exercise regularly but Orla does not and is not as fit.
Suggest **three** ways in which the data in the graph shows that Orla is less fit. [3]
(d) State **one** benefit of exercise on the heart. [1]
(e) Explain fully why the pulse rate increases during exercise. [4]

2.3 Reproduction, Fertility and Contraception

1. The diagram below represents the female reproductive system.

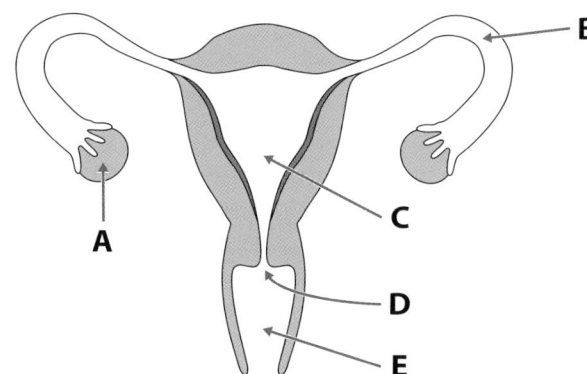

   (a) (i) Name structures **B**, **D** and **E**. [3]
       (ii) Give the functions of **A** and **C**. [2]
   (b) (i) State the letters, in sequence, of the parts of the female reproductive system through which sperm must travel if pregnancy is to occur. [1]
       (ii) Infections can sometimes lead to a blockage in one of structures labelled **B**. Suggest the effect of this on the ability of an affected female to become pregnant. Explain your answer. [3]

2. (a) Copy and complete the table below about structures and functions of the male reproductive system.

Structure	Function
prostate gland	
	the organ which introduces the sperm into the female vagina
scrotum	

   [3]

   (b) Sperm are specialised cells produced in the male reproductive system.
       (i) In terms of structures they pass through, describe the passage of sperm from where they are produced until they leave the male body. [4]
       (ii) The table below includes a structure of a sperm and the function of a structure. Copy and complete the table.

Structure	Function
haploid nucleus	
	structure which allows the sperm to swim through the female reproductive system to meet the egg

   [2]

UNIT 2: BODY SYSTEMS, GENETICS, MICROORGANISMS AND HEALTH

HT ONLY
HT ONLY
HT ONLY
HT ONLY

3. Sperm and the placenta are important components of the male and female reproductive systems respectively. They are each highly adapted for their functions.
   (a) Sperm cells possess numerous mitochondria. Explain the importance of this adaptation. [2]
   (b) The placenta contains villi. Describe and explain how these villi help the placenta to carry out its function. [3]

4. (a) Describe the events which take place in the female reproductive system between zygote formation and implantation in the uterus wall. [3]
   (b) The developing foetus is attached by the umbilical cord to the placenta.
       (i) Name **two** substances which pass along the umbilical cord to the foetus. [2]
       (ii) Describe the functions of the amnion and the amniotic fluid. [2]

5. (a) Testosterone causes secondary sexual characteristics to develop in males. Name **two** secondary sexual characteristics which develop in males. [2]
   (b) Name the hormone which causes the development of secondary sexual characteristics in females. [1]

6. (a) Menstruation occurs as part of the menstrual cycle.
       (i) Describe what is meant by the term menstruation. [1]
       (ii) Suggest why it is important that it takes place. [1]
   (b) The graph below shows how the levels of oestrogen and progesterone change during the menstrual cycle. Ovulation takes place around day 14.

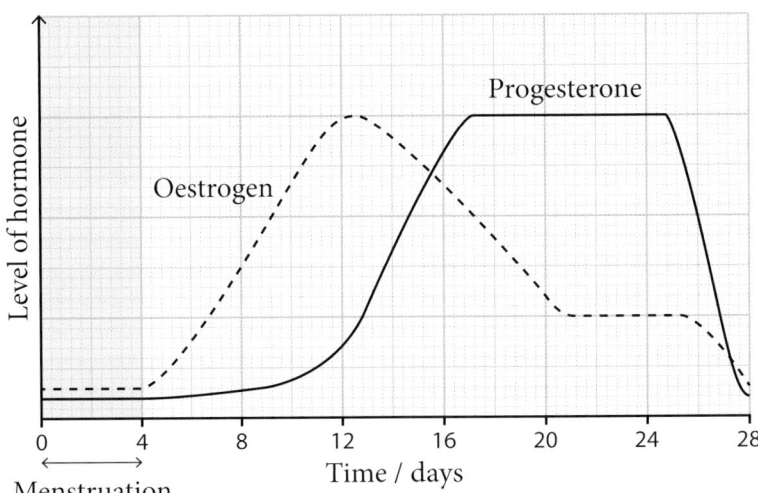

   (i) Using data from the table, describe how the level of progesterone changes during the menstrual cycle. [3]
   (ii) Give the evidence from the graph which suggests that oestrogen stimulates ovulation. [1]
   (iii) Give the function of progesterone between days 17–24. [1]
   (iv) Using the graph, suggest what causes menstruation to occur after day 28. [1]
   (v) Give **two** reasons why sex is more likely to lead to pregnancy on day 15 rather than day 5. [2]

2.3 REPRODUCTION, FERTILITY AND CONTRACEPTION

7. Some individuals are unable to have children. There are many possible reasons for infertility.
   (a) (i) Suggest **one** reason for infertility in males. [1]
   (ii) Suggest **one** reason for infertility in females. [1]
   (b) In vitro fertilisation is a form of treatment designed to overcome some forms of infertility. At an early stage of the treatment cycle the female is given hormones.
   (i) What is the function of this hormone treatment? [1]
   (ii) Describe the process of in vitro fertilisation. [3]
   (iii) Following in vitro fertilisation more than one embryo is usually transferred into the uterus. Explain the reason for this. [1]

HT ONLY

8. Contraception is the general term given to a range of techniques designed to prevent pregnancy.
   (a) Copy and complete the table below about types of contraception and how they work. Give **one** advantage and **one** disadvantage of each type.

Contraception	Method of action	Advantage	Disadvantage
(male) condom			
	prevents the ovaries from releasing eggs		
	cutting of oviducts prevents sperm from reaching the eggs		

[9]

(b) The bar chart below shows how many male sterilisation operations (vasectomies) took place for different age groups in a hospital over a one-year period.

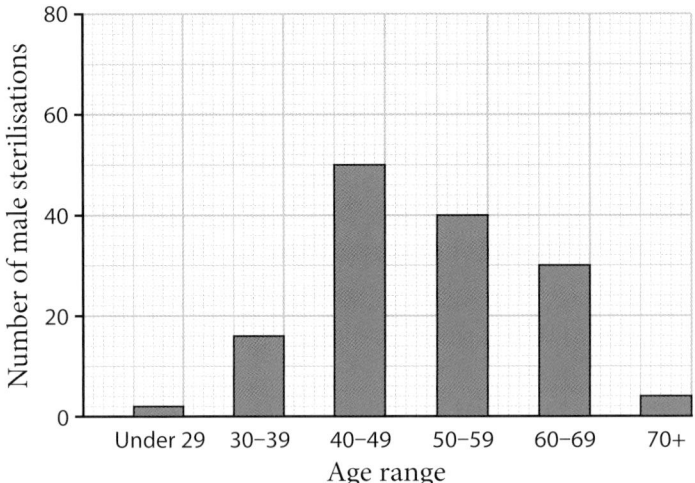

- **(i)** Give the mode for this data. [1]
- **(ii)** Suggest **one** reason for the relatively low numbers of males having vasectomies in the:
    - 20-29 age range;
    and the
    - 70 + age range. [2]
- **(iii)** Describe how a vasectomy prevents pregnancy. [2]

2.4 Genome, Chromosomes, DNA and Genetics

1. The diagram below represents a nucleus from an animal cell. One pair of chromosomes is shown.

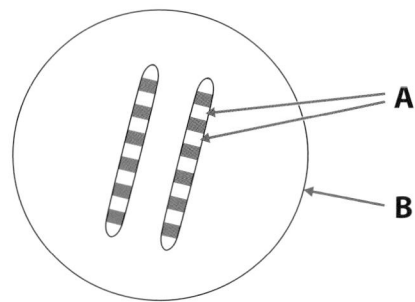

   (a) Name the structures labelled **A** and **B**. [2]
   (b) Name the part of an animal cell which lies immediately outside the nucleus. [1]
   (c) Give **one** reason why this cell could **not** be a bacterial cell. [1]

2. (a) The diagram below represents part of a DNA molecule.

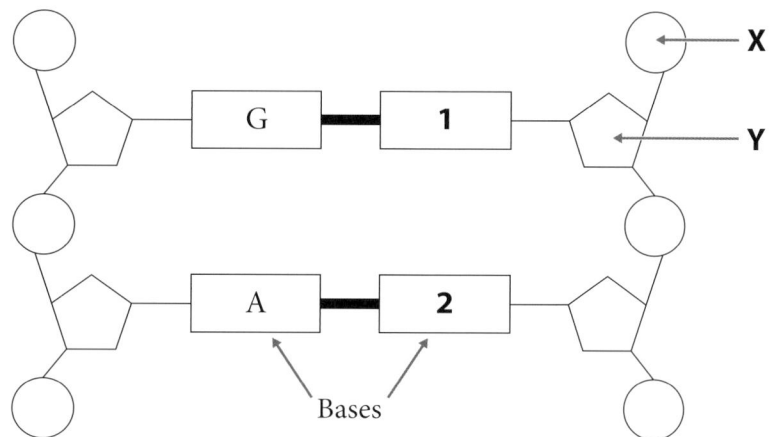

   (G = guanine; A = adenine)

   (i) Name structures **X** and **Y**. [2]
   (ii) Identify bases **1** and **2**. [2]
   (b) Twenty two percent of all the bases in an organism are guanine. Calculate the percentages for adenine, cytosine, and thymine. [3]
   (c) (i) Define the term genome. [1]
   (ii) Explain what is meant by the unique nature of an individual's DNA. [1]

3. The diagram below shows the role of DNA in building up amino acids in the correct sequence.

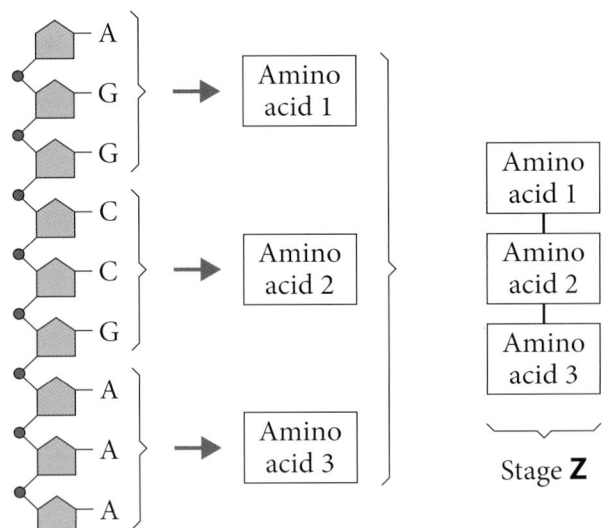

(a) Name the DNA base which is **not** shown in the diagram. [1]
(b) How many base triplets are shown in the diagram? [1]
(c) Using the diagram, describe what takes place during stage **Z**. [2]
(d) Calculate the number of bases required to code for a protein consisting of 126 amino acids. [2]

4. Mitosis and meiosis are types of cell division.
(a) Give **two** functions of mitosis in living organisms. [2]
(b) The diagram below summarises one type of cell division. Only two pairs of chromosomes are shown in each cell.

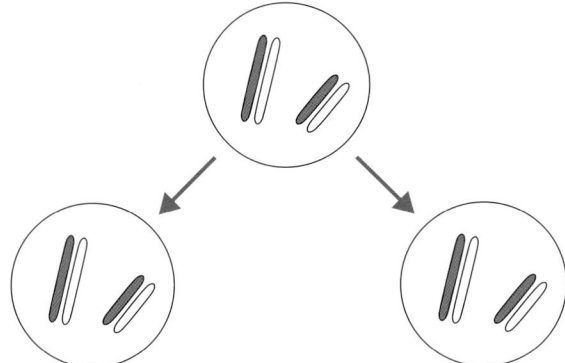

(i) Give **one** piece of evidence which indicates that the diagram summarises the process of mitosis. [1]
(ii) Using the diagram of mitosis above, describe **three** ways in which the process of meiosis is different from mitosis. [3]
(iii) State precisely where meiosis takes place in human males. [1]

2.4 GENOME, CHROMOSOMES, DNA AND GENETICS

5. **(a)** Define the following terms:
    - allele
    - homozygous
    - recessive allele [3]
    (b) In terms of sex chromosomes, distinguish between the chromosome arrangements in males and females. [2]

6. Flower colour in a type of pea can be either purple or white. Flower colour is determined by a single gene containing two different alleles. The allele for purple is dominant to the allele for white.

    Let **P** = purple; **p** = white.
    (a) A particular pea plant is heterozygous for flower colour:
       (i) Give this plant's genotype. [1]
       (ii) Give this plant's phenotype. [1]
    (b) This heterozygous plant was crossed with another heterozygous plant.
       (i) Draw and complete a Punnett square to show the parental gametes and the offspring genotypes. [3]
       (ii) What percentage of offspring would be expected to have purple flowers? [1]
       (iii) What percentage of offspring would be expected to be homozygous for flower colour? [1]
    (c) In another cross involving this type of pea, a plant with purple flowers was crossed with a plant which had white flowers. Fifty percent of the offspring of this cross had purple flowers and the other fifty percent had white flowers.
       (i) Draw and complete a Punnett square to show the parental gametes and the offspring genotypes. [3]
       (ii) How many different types of offspring genotypes were produced in this cross? [1]

7. Cystic fibrosis is an inherited disease. The allele for cystic fibrosis (**f**) is recessive to the allele **F**. Individuals who have an allele **F** will not have cystic fibrosis.
    (a) Draw and complete a Punnett square to show how it is possible for two parents who do not have cystic fibrosis to have a child who has the condition. [3]
    (b) In the Punnett square, circle the genotype for cystic fibrosis. [1]
    (c) What percentage of offspring from parents with these genotypes would be expected to have cystic fibrosis? [1]

UNIT 2: BODY SYSTEMS, GENETICS, MICROORGANISMS AND HEALTH

8. Phenylketonuria (PKU) is an inherited condition caused by a recessive allele. Individuals affected by PKU are unable to process a particular amino acid in the body. The pedigree diagram below shows the inheritance of PKU in a family.

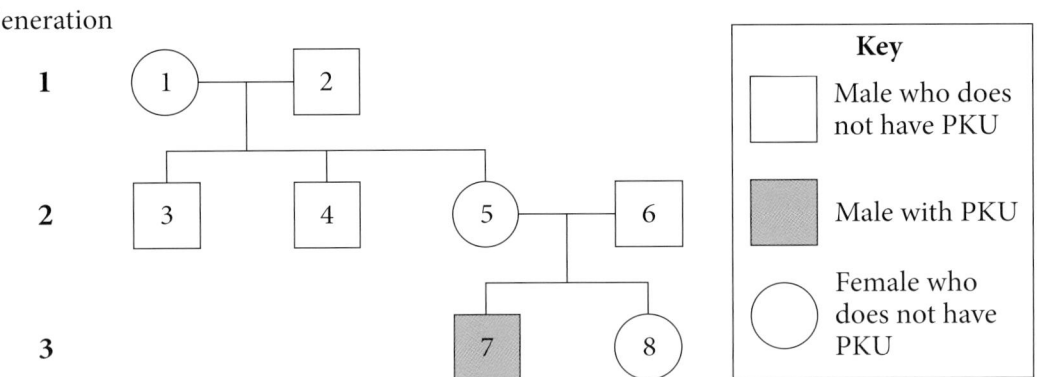

(a) What is the relationship of individual 3 to individual 2? [1]
(b) (i) Using the symbols **K** for the non-PKU allele and **k** for the PKU allele, give the genotype of individual 7. [1]
(ii) Identify the genotypes for individuals 5 and 6. [2]
(iii) If individuals 5 and 6 were to have another (third) child, what is the probability that this child will have PKU? [1]

9. Huntingtons's disease is caused by the presence of a dominant allele in a particular gene. Let **H** = Huntington allele; **h** = unaffected (normal) allele.
(a) (i) Give all the possible genotypes for individuals who have Huntington's disease. [1]
(ii) Give the genotype for individuals unaffected by Huntington's disease. [1]

(b) The pedigree diagram below shows the inheritance of Huntington's disease in a family.

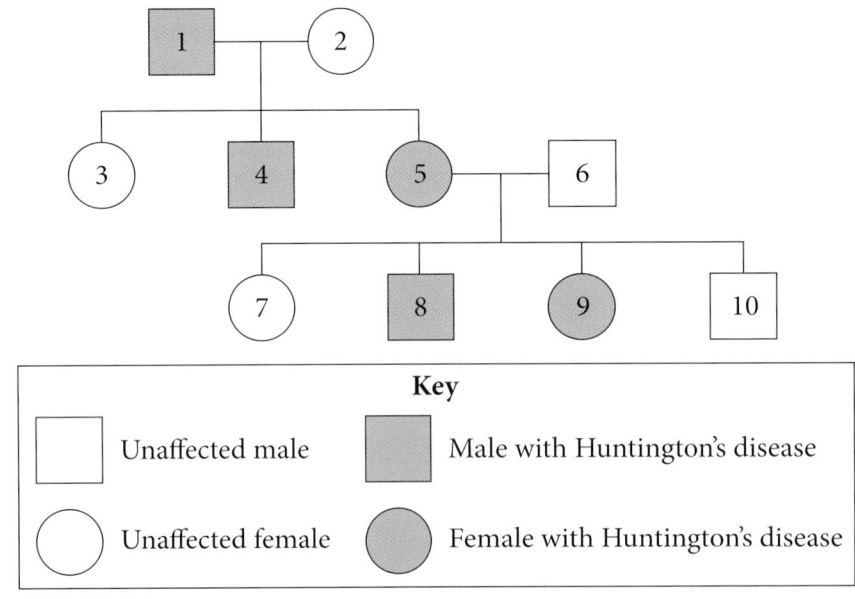

(i) What is the relationship of individual 3 to individual 4? [1]
(ii) Identify the genotypes of individuals 7 and 8. [2]
(iii) Calculate the probability of the next child of individuals 5 and 6 being a male who will develop Huntington's disease. [3]

2.4 GENOME, CHROMOSOMES, DNA AND GENETICS

10. (a) The diagram below shows how the gene for haemophilia is positioned on the sex chromosomes.

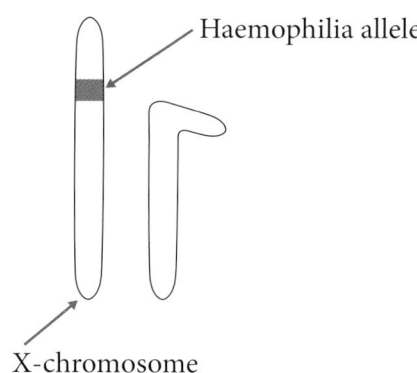

      (i) What name is given to the other chromosome in the diagram? [1]
      (ii) State the sex of the individual represented by the chromosomes shown. [1]

(b) The allele for haemophilia is recessive. The possible alleles used in genetic crosses involving the inheritance of haemophilia are usually represented as:

X^H; X^h; Y

The genotype $X^H X^h$ gives the carrier phenotype. Carriers are females who possess the haemophilia allele but do not have the condition.

      (i) Using a Punnett square, show how it is possible for two parents who do not have haemophilia to have a child with the condition. Identify the individual(s) in your cross who has/have haemophilia. [4]
      (ii) Haemophilia is described as a sex-linked condition. Suggest what is meant by the term sex-linked. [1]

11. (a) Down's Syndrome is a condition caused by errors in gamete formation. The condition arises when a gamete from one parent has an extra chromosome and has 24 chromosomes. If this gamete is involved in fertilisation, give the number of chromosomes that will be present in the zygote. [1]

(b) Down's Syndrome can be detected using amniocentesis.
      (i) Outline the steps involved in an amniocentesis test. [3]
      (ii) Some individuals offered an amniocentesis test decide against having this procedure. Suggest **one** reason why. [1]

12. Huntington's disease is an inherited condition which usually only becomes apparent in affected individuals when they are around forty years old. Huntington's disease leads to the breakdown of nerve tissue over time and there is no cure. It is caused by a dominant allele.

(a) What is the probability of an individual developing Huntington's disease if they have one parent heterozygous for the condition and one parent who does not have the allele for the condition? Show your working out by using a Punnett square to explain the genetic cross involved.

Let **H** = Huntington allele; **h** = unaffected (normal) allele. [4]

(b) It is possible to do a test in younger people to show if an individual will develop Huntington's disease when they are older.
  (i) State the name which is used to describe testing individuals for the likelihood of developing an inherited disease. [1]
  (ii) Suggest **one** reason why a female in her early twenties who has a parent with Huntington's disease might decide **not** to have the test to show if she will develop the condition when she is older. [1]
(c) Many individuals take out life insurance policies. These policies can provide finance to the families of individuals should they die or become very ill. There has been considerable debate as to whether genetic information about individuals should be made available to the insurance companies that provide these policies. Give **one** argument for **not** making genetic information available to the insurance companies. Explain your answer. [2]

13. The DNA of bacteria can be modified in such a way that the bacteria will produce human insulin. To be able to produce human insulin the bacteria must have the human insulin gene in its DNA. This involves several steps. One of the steps is summarised in the diagram below.

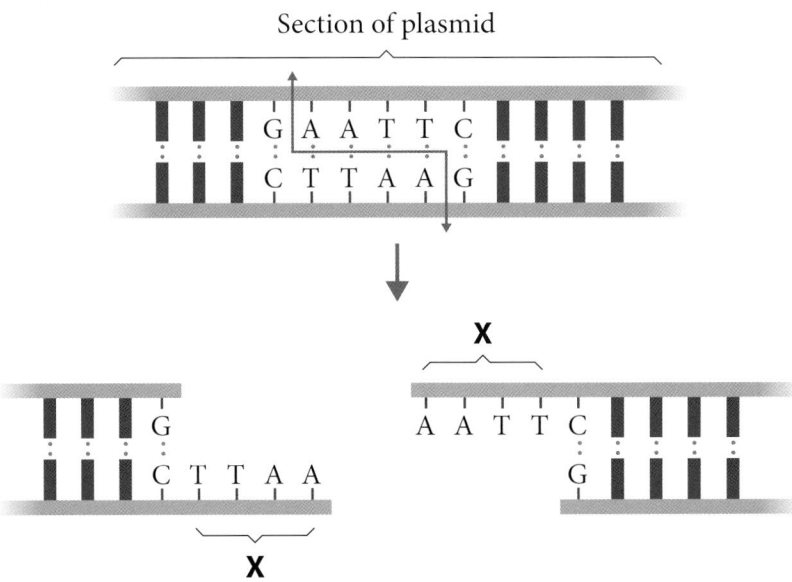

(a) Name the type of enzyme used to cut DNA as shown in the diagram. [1]
(b) What name is used to describe the sections labelled **X**? [1]
(c) Using the diagram and your knowledge, describe the sequence of events which take place following the cutting of the plasmid. [3]

14. Human insulin can be made using genetic engineering. The process involves introducing the human insulin gene into the genome of bacteria. Following this, the bacteria produce human insulin.

(a) Describe the processes involved in producing bacteria which contain the human insulin gene. Your account should include the role of restriction enzymes and bacterial plasmids.

In this question, you will be assessed on your written communication skills including the use of specialist scientific terms. [6]

(b) Before the advent of genetic engineering to produce insulin, the insulin used to treat people with diabetes had to be extracted from the pancreases of domestic animals such as pigs. Give **two** advantages of producing insulin by genetic engineering. [2]

2.5 Variation and Natural Selection

1. **(a)** Humans have four major blood groups (A, B, AB, and O). This is an example of discontinuous variation in human populations. One hundred people were tested to find their blood group and the results are shown in the table below.

Blood group	Number of people
A	42
B	10
AB	8
O	40

   (i) Define the term 'discontinuous variation'. [1]
   (ii) Name the type of graph that should be used to present this type of data. [1]
   (iii) On graph paper, draw an appropriate graph to display this blood group data. [4]

   (b) Apart from blood groups, give **one** other example of discontinuous variation in humans. [1]

2. **(a)** The table below gives information about some features in four teenagers.

Teenager	Weight / kg	Tongue roller	Hand dominance	Height / cm
Rick	58	yes	right	172
Máire	47	yes	right	149
Jane	42	no	right	153
Tony	55	yes	left	178

   (i) Which **two** features in the table show continuous variation? [2]
   (ii) Calculate the mean weight of the four teenagers. Give your answer to one decimal place. [2]

(b) In a separate investigation, the heights of all the students in a class were measured. The results are shown in the graph below.

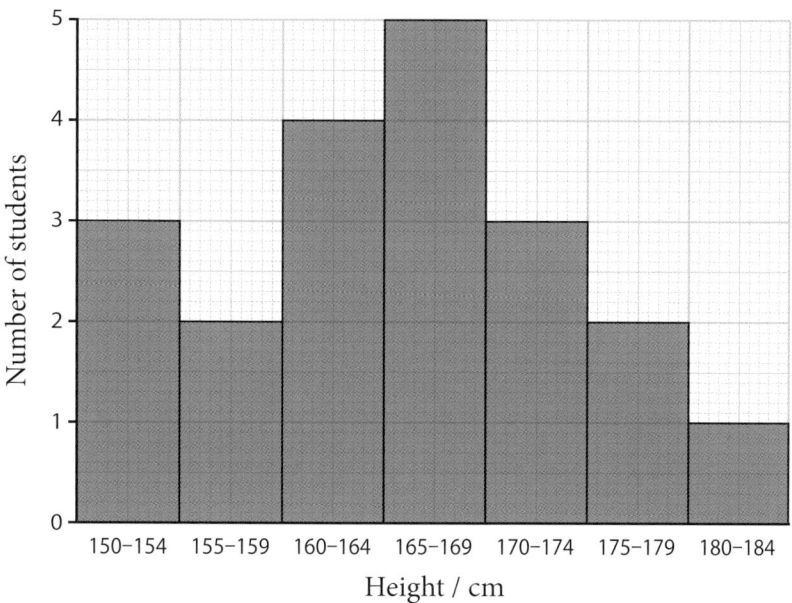

- **(i)** Name the type of graph used to present the data of student height. [1]
- **(ii)** How many students are in this class? [1]
- **(iii)** What percentage of students in the class had a height in the range 160–174 cm? [3]
- **(iv)** State the type of variation shown by the data in the graph and explain the evidence for this. [2]

3. Lesser Celandine is a common woodland plant which usually has between 6 and 12 petals in each of its flowers. The photograph below shows a flower of Lesser Celandine.

The number of petals in Lesser Celandine flowers growing in woodland was counted and the results are shown in the graph below.

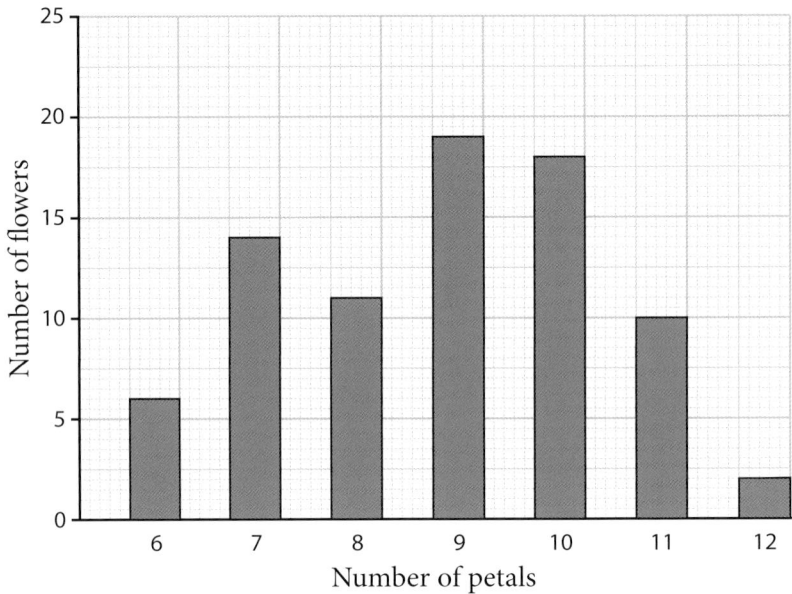

(a) Name the type of graph shown. [1]
(b) Name the type of variation shown by petal number in Lesser Celandine. [1]
(c) What percentage of flowers had the same number of petals as the flower in the photograph? Give your answer to one decimal place. [3]

4. Variation in living organisms can have a genetic or environmental basis or be a result of both.
   (a) Genetic variation can be due to mutations or sexual reproduction.
       (i) Explain what is meant by mutation. [1]
       (ii) Explain how sexual reproduction leads to variation. [2]
   (b) Explain how height in humans can be affected by both genetics and the environment. [2]

5. The peppered moth is a moth which can be found on the trunks of trees. This moth has two genetically distinct forms – a lighter grey 'light' variety and a much darker 'dark' form. Both varieties of peppered moth are a food source for several species of birds. Moths which are better camouflaged when on the tree trunk are less likely to be spotted by birds.

In industrial or urban areas, the trees are often blackened due to smoke whereas in unpolluted areas the tree bark retains its natural lighter colour.

The graph below shows how the level of smoke pollution in an area changed over a period of 30 years.

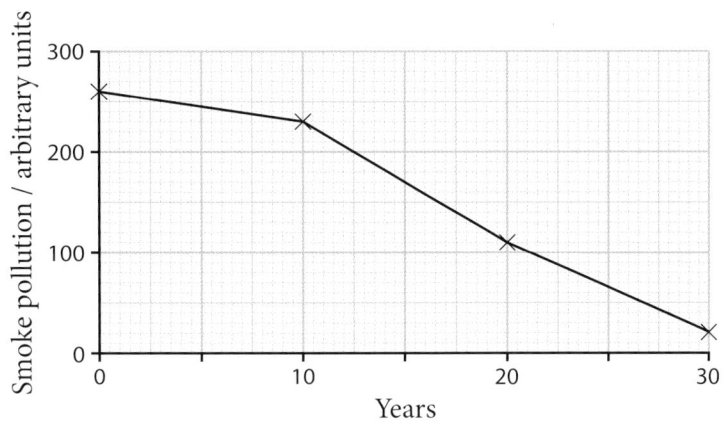

(a) Describe the trend for smoke pollution over the 30 years. [1]

(b) Every 10 years over the 30-year period the numbers of each type of peppered moth were estimated in the same area where the smoke pollution was monitored. The results are shown in the table below.

Year	Number of peppered moths	
	Light variety	Dark variety
0	121	489
10	146	450
20	473	128
30	569	37

Using all the information provided and your understanding of natural selection, describe and explain the results shown in the table. [5]

6. Most types of grass are unable to grow on soils which are contaminated by zinc, lead or copper. This type of 'heavy' metal contamination is common in the soils of areas where these metals have been mined in the past.

Some types of grass have a mutation which allows them to grow on contaminated soil (they are tolerant of heavy metals). However, the mutation causes the grass plants to grow at a much slower rate than grass plants which do not have the mutation.

The graph below shows how the percentages of grasses which are tolerant of heavy metal change as the level of soil contamination changes.

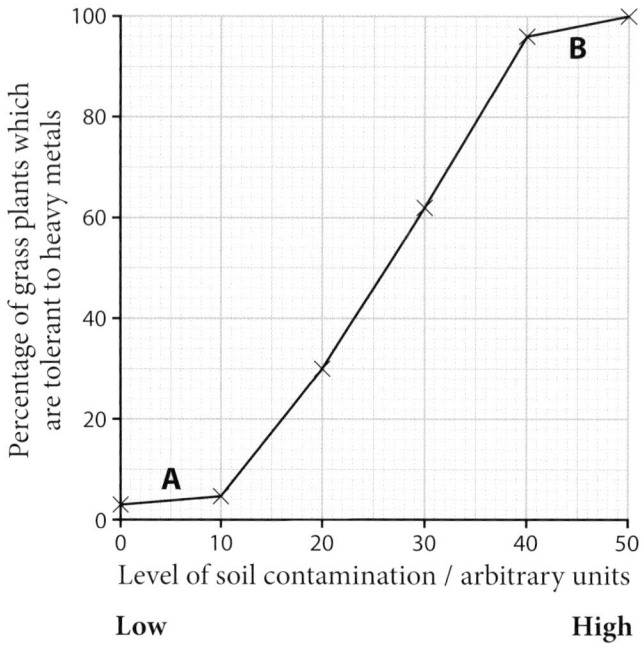

(a) Describe the main trend shown by the graph. [1]
(b) Using the information provided and your knowledge of competition for resources and natural selection, explain the results at positions **A** and **B** on the graph. [4]

7. (a) The table below shows the total number of species extinctions which have occurred in a country between 1960 and 2010. Each value represents the total number of extinctions which have taken place by the year shown.

Year	Total number of extinctions
1960	3
1970	6
1980	10
1990	22
2000	31
2010	54

2.5 VARIATION AND NATURAL SELECTION

      (i) Draw a line graph to represent the information in the table. [4]
      (ii) Predict how many extinctions there will be by the year 2020. [1]
  (b) (i) Explain what is meant by the term extinction. [1]
      (ii) Suggest **one** reason why species become extinct. [1]

8. On one small island close to the coast of America, many species of insects are wingless. The island is affected by very strong winds. Research has shown that insects that do have wings are at risk of being blown out to sea and dying. *HT ONLY*

One species of wingless insect was investigated in detail. Analysis has shown that in the distant past, both winged and wingless forms existed, but that now only the wingless variety exists.
  (a) What is the evidence that evolution has taken place in this species? [1]
  (b) Explain how variation in phenotypes and natural selection have contributed to change in this species. [3]
  (c) Explain how fossils provide evidence of evolution. [2]

9. Natural selection and selective breeding both lead to changes in species over time.
  (a) Explain the difference between selective breeding and natural selection. [2]
  (b) Some types of cattle are commercially important because of their milk production. These types of cattle have been subjected to selective breeding over a long period of time to maximise milk production. Describe the process of selective breeding which has resulted in cattle that can produce large volumes of milk. [4]

2.6 Health, Disease, Defence Mechanisms and Treatments

1. **(a)** The table below shows some information about four diseases caused by bacteria or viruses.

Microorganism involved	Bacteria or virus	How spread
Salmonella	Bacteria	
HIV		Sexual contact
Flu		
Tuberculosis		Airborne (droplet) infection

       (i) Copy and complete the table to include the missing information. [5]
       (ii) Describe **one** way in which it is possible to prevent being affected by salmonella poisoning. [1]
       (iii) Diseases such as flu and HIV are described as communicable. Describe what is meant by the term communicable. [1]
   (b) Name **one** human disease caused by a fungal infection. [1]

2. **(a)** Blood clotting is important in preventing the loss of blood but is also important in defence against disease.
   Explain how blood clotting following a skin wound helps protect the body against disease. [2]
   (b) Antibodies are also important in defence against disease.
       (i) Name the type of white blood cell which produces antibodies. [1]

   The diagram below represents a disease-causing microorganism and an antibody.

   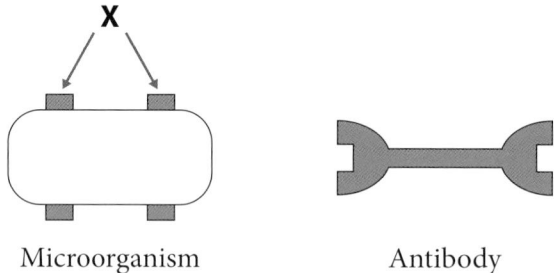

       (ii) Identify structure **X**. [1]
       (iii) Using the diagram, explain how antibodies help protect against disease. [3]

The diagram below represents a different microorganism.

      (iv) Describe how the body would need to react to have a successful antibody reaction against this microorganism. [2]
- **(c)** Antibodies can be involved in both active and passive immune responses. Explain the difference between active and passive immunity. [2]
- **(d)** Phagocytes are another type of white blood cell important in defence against disease. Describe the role of phagocytes in defence. [2]

3. (a) The graph below shows how antibody level changes following an infection caused by a virus.

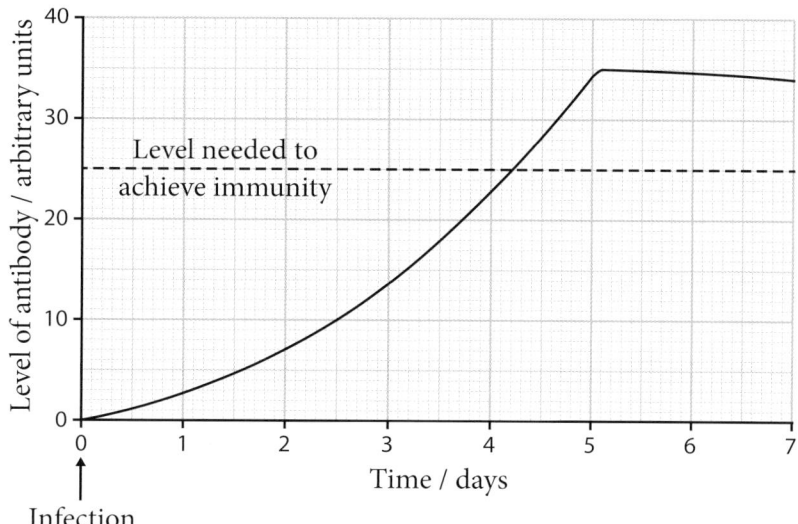

      (i) How many days after the infection did it take for immunity to be achieved? [1]
      (ii) Suggest the reason for this delay. [1]
      (iii) Immunity can be passive or active. Name the type of immunity shown by the graph and give **one** piece of evidence to support your answer. [2]
- **(b)** Memory lymphocytes are important in providing long term protection against disease.
      (i) What causes the development of memory lymphocytes? [1]
      (ii) Explain why they are effective in protecting against disease. [1]

UNIT 2: BODY SYSTEMS, GENETICS, MICROORGANISMS AND HEALTH

4. Vaccinations typically contain dead or modified microorganisms which can be injected into people.
   (a) Explain why the microorganisms in a vaccination must be dead or modified. [1]

   HT ONLY
   (b) The graph below shows how vaccination can affect antibody level in the body.

   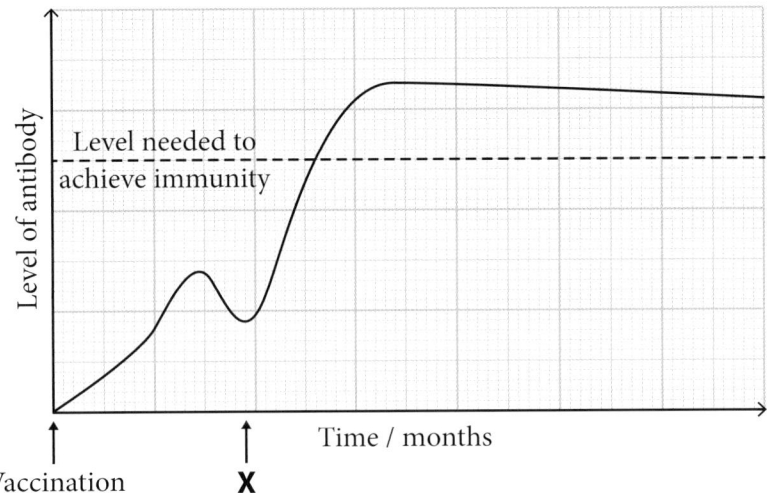

   (i) Suggest what **X** represents. [1]
   (ii) Explain how vaccinations help provide long-term immunity. [2]

   (c) Colds and flu are caused by viruses. These viruses are relatively unstable and have high rates of mutation leading to many different types or strains of cold and flu. Colds are usually much more common than flu. Colds are also much less harmful to those affected as flu can make people very unwell.

   There are vaccinations for flu but not for colds. Not everyone gets a flu vaccination, but they are offered to older people above a certain age.

   Using the information provided and your knowledge, answer the questions below.
   (i) Suggest why older people are offered a flu vaccination but younger healthy people are not. [2]
   (ii) Suggest why a different flu vaccination is provided each year. [1]
   (iii) Suggest why there is not a vaccination for the cold. [1]

5. (a) The diagram below represents a small section from the top surface of a leaf.

   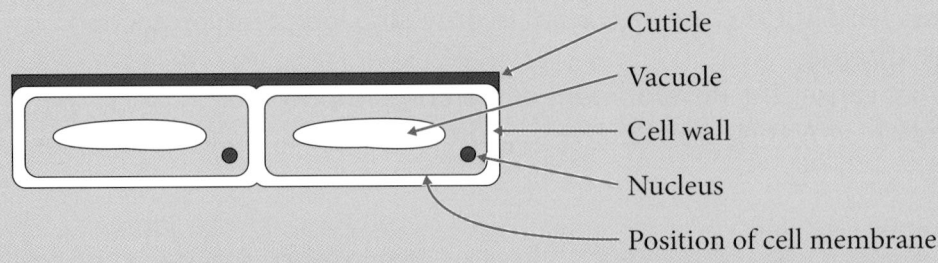

   (i) Name the **two** labelled structures important in restricting the entry of microorganisms into the leaf. [2]
   (ii) For **one** of these structures, suggest how it is adapted to restrict microorganism entry. [1]

(b) Many trees and plants produce poisonous chemicals, particularly in parts which are damaged. For example, Cherry Laurel produces cyanide and Foxglove produces digitalis.
  (i) Give the function of these chemicals. [1]
  (ii) Suggest why these chemicals are more likely to be produced in greater numbers at parts of a plant which become damaged. [1]

6. (a) Explain what is meant by the term 'superbug'. [1]
   (b) In an investigation, bacteria were cultured in nutrient broth in a beaker. The estimated numbers of bacteria over a six-day period were calculated. At one point during the six-day period an antibiotic was added to the broth. The results are shown in the graph below.

   (i) During which day was the antibiotic added? [1]
   (ii) Suggest **two** possible reasons why the numbers of bacteria started to rise from day 4. [2]
   (c) Superbugs are very difficult to eradicate and a range of measures are used to limit their spread. Both scientists and the public can help eliminate or reduce the spread of superbugs.
   (i) Suggest **one** way in which scientists can help eliminate or reduce the spread of superbugs. [1]
   (ii) Suggest **one** way in which individual members of the public can help reduce the spread of superbugs. [1]

7. (a) Tobacco smoke contains a range of harmful substances. These include tar, nicotine and carbon monoxide.
   (i) Individuals who smoke tobacco are at a higher risk of developing bronchitis. Describe what bronchitis is and name the product in tobacco smoke which causes it. [2]
   (ii) Nicotine can affect the heart rate. Give **one** other effect it has. [1]
   (iii) What is the effect of carbon monoxide on the oxygen-carrying capacity of the blood? Explain your answer. [3]
   (iv) Describe the effect of emphysema on the lungs. [2]

(b) Individuals who have been smoking tobacco for a long period of time are often very short of breath and lack energy. Using the information provided and your knowledge, explain this shortage of breath and lack of energy. [5]

(c) Electronic cigarettes (E-cigarettes) have become increasingly popular in recent years. Unlike traditional cigarettes, they do not contain tobacco and tar. However, most contain nicotine and a wide range of chemicals, with the chemicals being in a lower concentration than in traditional tobacco-containing cigarettes.

In an investigation, the numbers of individuals who smoked each of traditional cigarettes and E-cigarettes in a small town were estimated over a twelve-year period. The results are shown in the graph below.

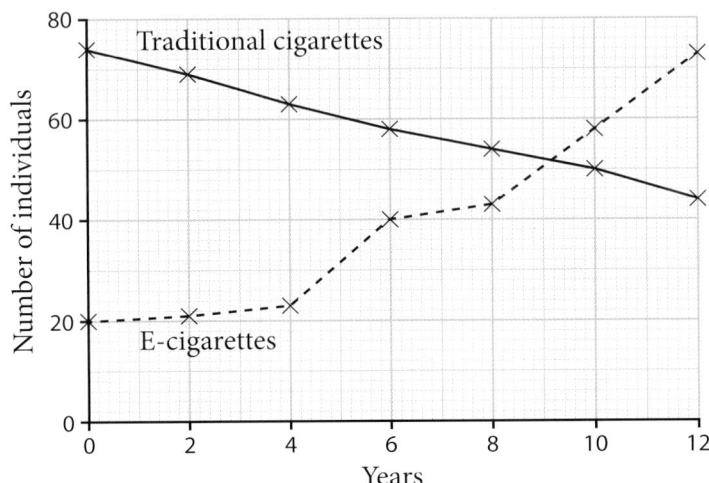

It has been suggested that the rise in E-cigarette use is caused by people switching from using traditional cigarettes to E-cigarettes. Using the graph and the information provided, answer the questions below.

(i) Give **one** piece of evidence from the graph to support the suggestion that the rise in E-cigarette use is caused by people switching from traditional cigarettes to E-cigarettes. [1]

(ii) State **one** piece of evidence which shows that the rise in people using E-cigarettes is not wholly caused by people switching from traditional cigarettes. [1]

(iii) Suggest **one** advantage of individuals giving up traditional cigarettes and switching to E-cigarettes. [1]

(iv) Suggest **one** reason why many health professionals advise young people who have never smoked either type of cigarette **not** to start using E-cigarettes. [1]

2.6 HEALTH, DISEASE, DEFENCE MECHANISMS AND TREATMENTS

8. The table below shows mean cholesterol levels for men and women of different age groups in a local area.

Age group	Men	Women
21–30	3.9	3.7
31–40	4.7	4.6
41–50	5.1	4.8
51–60	5.4	5.0
61+	5.8	5.1

   (a) Give **two** conclusions that can be made from the data in the table. [2]
   (b) (i) Explain how too much cholesterol in the diet can lead to restricted blood flow and eventually clot formation. [2]
       (ii) Give the function of the coronary arteries. [1]
       (iii) Suggest why clot formation is more likely to occur in the coronary arteries than in other arteries in the body. [1]
   (c) (i) Explain fully what causes a heart attack. [3]
       (ii) Give **two** lifestyle factors which will help reduce heart disease. [2]
   (d) One form of treatment for heart disease is the use of stents. Describe what a stent is and explain how it reduces the risk of a heart attack. [3]

9. There are many different types of cancer, and they have many causes.
   (a) (i) Describe the cause of skin cancer. [2]
       (ii) Suggest **one** reason why the number of cases of skin cancer is increasing. [1]

   Cancer is uncontrolled cell division. Cancer cells often grow faster and divide faster than normal cells. Cancer cells are often concentrated into a clump.

   (b) State the name that is normally used to describe a 'clump' of cancer cells. [1]
   (c) The diagram below represents a growth of cancer cells at the edge of a blood vessel.

   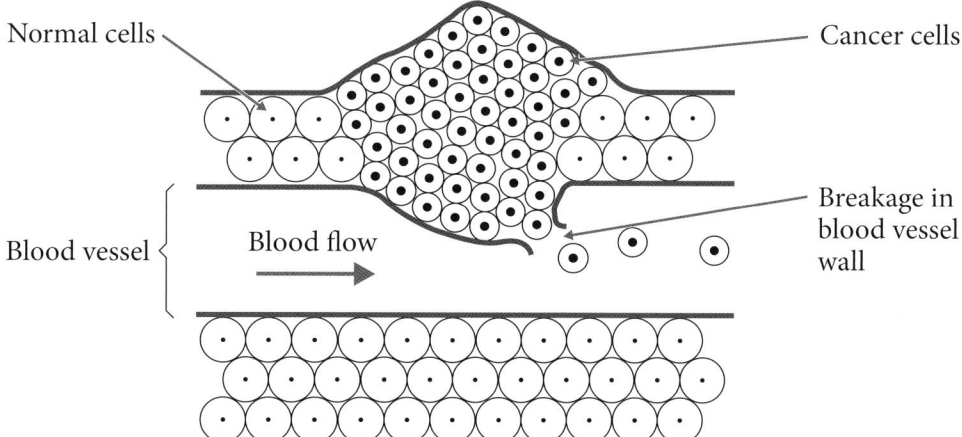

       (i) Suggest **one** reason why the cancer cells are smaller than normal cells. [1]
       (ii) What is the evidence from the diagram which suggests that this is a malignant growth? [1]
       (iii) Suggest **one** reason why malignant cancer is more difficult to treat by surgery than benign growth. [1]

UNIT 2: BODY SYSTEMS, GENETICS, MICROORGANISMS AND HEALTH

(d) There is considerable research into the treatment of cancer. New drugs or new treatments developed by scientists are subject to peer review before they are used in treatments.
- **(i)** Explain what is meant by the term peer review. [2]
- **(ii)** Explain why peer review is important. [1]

10. Prostate cancer is a cancer which develops in the prostate gland of men. It is the commonest male cancer in the UK. Approximately one in eight men will get prostate cancer at some stage of their lives.
- **(a)** Assuming that there are equal numbers of men and women, calculate the number of individuals who would be expected to get prostate cancer in a population of 5472. [2]
- **(b)** Men are more likely to develop prostate cancer if a close relative, e.g. father or brother, has the condition. Suggest **one** reason for this. [1]
- **(c)** Many countries offer a screening programme for prostate cancer.
  - **(i)** Explain why this screening programme is offered to men and not women. [1]
  - **(ii)** Suggest **one** reason why this screening programme is particularly targeted at older men (50+). [1]
  - **(iii)** In terms of health benefits, explain the benefits of cancer screening programmes. [2]
  - **(iv)** Some men are reluctant to take part in cancer screening programmes. Suggest **one** reason for this. [1]

11. (a) Outline Fleming's role in the discovery of penicillin. [2]
- **(b)** New medicines and drugs must go through several stages before they can be used to treat members of the public. An early stage in drug development involves preclinical trials. This includes testing the drugs on cells and tissues in the laboratory. This is typically a 'trial and error' exercise, with only a minority of the drugs tested proceeding beyond this stage.
  - **(i)** Suggest **one** reason for testing drugs on cells and tissues in the laboratory. [1]
  - **(ii)** Following testing on cells and tissues, the drugs are typically tested on animals. Suggest **one** reason for testing drugs on animals. [1]
- **(c)** The final stage in drug development involves clinical trials.
  - **(i)** Describe what is meant by a clinical trial. [1]
  - **(ii)** Give **one** reason for carrying out clinical trials. [1]
- **(d)** Sometimes drugs may be effective in treating a disease but have side effects. A side effect is an unwanted effect a drug may have. For example, many patients receiving chemotherapy as a cancer treatment lose their hair and suffer from nausea (feeling sick). Using the information provided, suggest why a drug could be approved as a treatment even though it has side effects. [1]
- **(e)** Using the information provided, suggest **one** reason why the development of a drug is normally very expensive. [1]

12. An investigation was carried out to see the effects of three different antibiotics on bacteria. Nutrient agar in a Petri dish was inoculated by a strain of bacteria to create a lawn of bacteria.

(a) Suggest what is meant by the term 'lawn' of bacteria. [1]

Three paper disks (**A**, **B** and **C**) were soaked in different antibiotics and added to the Petri dish. The Petri dish was incubated at 25 °C for 48 hours. The results are shown below.

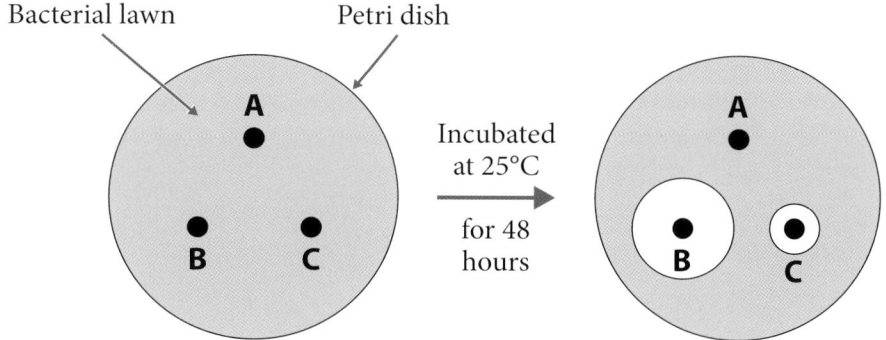

(b) Describe and explain the results. [3]
(c) State **one** variable that would need to be controlled in this investigation. [1]
(d) It is important that appropriate aseptic techniques are used in this investigation.
   (i) Explain why Petri dishes are only partially opened when adding the bacteria. [1]
   (ii) Describe and explain **one** other aseptic technique relevant to this investigation. [2]
(e) Explain why the Petri dish was incubated at 25 °C rather than 35 °C. [2]